Digital Design and Implementation with Field Programmable Devices

Digital Design and Implementation with Field Programmable Devices

Zainalabedin Navabi
Northeastern University

Distributors for North, Central and South America:
Kluwer Academic Publishers
101 Philip Drive
Assinippi Park
Norwell, Massachusetts 02061 USA
Telephone (781) 871-6600
Fax (781) 871-6528
E-Mail: < kluwer@wkap.com >

Distributors for all other countries:
Kluwer Academic Publishers Group
Post Office Box 322
3300 AH Dordrecht, THE NETHERLANDS
Telephone 31 78 6576 000
Fax 31 78 6576 254
E-Mail: < services@wkap.nl >

 Electronic Services < http://www.wkap.nl >

Library of Congress Cataloging-in-Publication Data

Title: Digital Design and Implementation with Field Programmable Devices
Author(s): Zainalabedin Navabi
ISBN: 1-4020-8011-5
ISBN: 1-4020-8012-3 (eBook)

Dr. Zainalabedin Navabi is an adjunct professor of electrical and computer engineering at Northeastern University. Dr. Navabi is the author of several textbooks and computer based trainings on VHDL, Verilog and related tools and environments. Dr. Navabi's involvement with hardware description languages begins in 1976, when he started the development of a register-transfer level simulator for one of the very first HDLs. In 1981 he completed the development of a synthesis tool that generated MOS layout from an RTL description. Since 1981, Dr. Navabi has been involved in the design, definition and implementation of Hardware Description Languages. He has written numerous papers on the application of HDLs in simulation, synthesis and test of digital systems. He started one of the first full HDL courses at Northeastern University in 1990. Since then he has conducted many short courses and tutorials on this subject in the United States and abroad. In addition to being a professor, he is also a consultant to CAE companies. Dr. Navabi received his M.S. and Ph.D. from the University of Arizona in 1978 and 1981, and his B.S. from the University of Texas at Austin in 1975. He is a senior member of IEEE, a member of IEEE Computer Society, member of ASEE, and ACM.

To my wife, Irma and my sons Arash and Arvand.

CONTENTS

This book is on digital system design for programmable devices, such as FPGAs, CPLDs, and PALs. A designer wanting to design with programmable devices must understand digital system design at the RT (Register Transfer) level, circuitry and programming of programmable devices, digital design methodologies, use of hardware description languages in design, design tools and environments; and finally, such a designer must be familiar with one or several digital design tools and environments. Books on these topics are many, and they cover individual design topics with very general approaches. The number of books a designer needs to gather the necessary information for a practical knowledge of design with field programmable devices can easily reach five or six, much of which is on theoretical concepts that are not directly applicable to RT level design with programmable devices.

The focus of this book is on a practical knowledge of digital system design for programmable devices. The book covers all necessary topics under one cover, and covers each topic just enough that is actually used by an advanced digital designer. In the three parts of the book, we cover digital system design concepts, use of tools, and systematic design of digital systems.

In the first chapter, design methodologies, use of simulation and synthesis tools and programming programmable devices are discussed. Based on this automated design methodology, the next four chapters present the necessary background for logic design, the Verilog language, programmable devices, and computer architectures.

Presenting design and use of design tools based on the methodology discussed in the first part of the book becomes meaningful, only if a real industrial tool is used. For this purpose, the second part of the book presents design of small components using simulation, synthesis and design entry tools provided by Altera's Quartus II design environment. While practicing design methodology of the first part of the book, this part familiarizes readers with the use of Quartus II integrated design environment.

The third part of the book discusses RT level system design. A top-down systematic approach is presented for design of relatively complex systems. This part shows how a design is partitioned into its lower-level components, how synthesis tools or predefined parts are used for implementation of RT level components, and how a complete system is put together and used for programming a programmable device.

The book can be used by hardware design practitioners who are already familiar with basics of logic design and want to move into the arena of automated design and design implementation using filed programmable devices. For this audience, this book provides a recap of digital design topics and computer architectures and shows the Verilog language for synthesis. In addition, for an industrial setting, the book shows how existing design components are used in upper level designs, and how user libraries are formed and utilized. Using Altera's UP2 programmable device development board with this book helps engineers test and debug their designs before programming their programmable devices on production boards.

In an educational setting, the book can be used as a complementary book for the basic logic design course, or a laboratory book for the sophomore logic design lab, or as a textbook for senior level design courses. Using Altera's UP2 programmable device education board with this book helps students see their designs being implemented and tested, and thereby get a down-to-wire understanding of how things work. For students in other fields of engineering like mechanical and chemical engineering, the book is a useful tool for design and implementation of controllers and interfaces.

OVERVIEW OF THE CHAPTERS

An overview of the chapters is given here. The first five chapters cover the main concepts of digital design with field programmable devices from a practical point of view. The next part of the book, in five chapters, shows the use of Altera's Quartus II as a typical FPLD design environment. The last four chapters cover complete digital designs that utilize various tools and utilities provided by a design environment like Quartus II.

Chapter 1 discusses the general flow of a digital design using tools available in design environments. This chapter is introductory and introduces tools and design methodologies.

Chapter 2 discusses basic logic design from a practical point of view. Only topics used for an automated design are discussed here.

Chapter 3 introduces Verilog. Synthesizable Verilog is emphasized, but for a complete HDL based design, testbenches and language utilities for this purpose are also discussed.

Chapter 4 talks about programmable devices. The approach we take is showing how original ROMs evolved into today's complex FPGAs.

In Chapter 5 we talk about digital design architectures. We show the basics of CPU architecture and how one goes about designing a processor.

Chapter 6 of this book discussed tools we use for design validation, synthesis, device programming and prototyping. We discuss the use of Quartus II, ModelSim HDL simulator and the UP2 development board.

Chapter 7 shows basic schematic entry for gate level designs. In this chapter we show the use of Quartus or simulation and device programming.

Chapter 8 shows the formation of a design library by developing commonly used parts, testing them and making available in a user library.

In Chapter 9, we show how parts from a user library and configurable parts from a design library can be put together for generating a complete design.

Chapter 10 shows HDL based design, simulation, synthesis and device programming. Only the synthesizable subset of Verilog is used for the design of this chapter.

Chapter 11 that is the first of the four complete designs of this book shows the design of a sequential multiplier by partitioning it into a data and a control part. Top-down design with Verilog is shown here.

In Chapter 12 a VGA interface is designed. We show how Verilog, gate level schematics, configurable library parts, and definable memories can be mixed in a complete design. In addition, the operating of a VGA monitor is discussed here.

A keyboard interface is designed in Chapter 13. In addition to showing the operation of a keyboard, we show a design that consists of schematics and HDL entry.

The CPU of Chapter 14 is a complete CPU that is primarily designed with Verilog. Testing of this CPU in Verilog and use of high-level test related tasks are discussed here.

ACKNOWLEGEMENTS

Several people helped me with preparation of this manuscript. My former student Mr. Saeed Safari wrote the chapter on computer architectures. He developed the example presented in this chapter and presented the design procedure using his example. Mr. Aryan Navabi who is a freshman in Computer Engineering did all the artwork in this book. His thoroughness and emphasis on the details were useful in generation of descriptive diagrams of the book. As with all my other publishing work, Ms. Fatemeh Asgari helped me with the preparation of the manuscript. She helped me with the organization of this work and allocation of time to this and many other projects I am involved in.

Instrumental in the original proposal and arrangement of this book was Mr. Mike Phipps of Altera. His guidelines in making this book useful for students and practitioners were helpful in the organization of the book. I thank him for his support and special attention to computer engineering education.

I also thank my wife, Irma Navabi, for help encouragement and understanding of my working habits. Such an intensive work could not be done if I did not have the support of my wife and my two sons, Arash and Arvand. I thank them for this and other scientific achievements I have had.

Zainalabedin Navabi
Boston, Massachusetts
April, 2004

Part
1

Digital Design Concepts

This part provides a practical knowledge of logic design concepts. The focus is on those digital design topics that are necessary for design and implementation of programmable logic devices using design automation tools and environments. Topics covered here are:

- Programmable Logic Based Design
- Digital Logic
- Practical Verilog
- Programmable Logic Devices
- Computer Architecture Design

1
PLD Based Design

This chapter presents tools and environments that are used for design with Field Programmable Logic Devices. We discuss steps involved in taking a hierarchical, high-level design from a description of the design to its implementation in an FPLD. Processes and terminologies are illustrated here. After the first section that discusses design flow, the proceeding sections elaborate on each step of this design flow.

1.1 Design Flow

For the design of FPLDs, the design flow begins with specification of the design and ends with programming the target device. Figure 1.1 shows steps involved in this design flow.

In the design entry phase, a design is specified as a mixture of block diagram and textual specifications. After performing pre-synthesis simulation, this design is taken through the synthesis process to translate it into actual hardware of the target device. Here, target device refers to the FPLD that is being programmed for the implementation of our design. After the synthesis process and before the actual device is programmed, another simulation is done that is referred to as, post-synthesis simulation. The difference between pre- and post-synthesis simulations is in the level of details obtained from each simulation.

The sections that follow elaborate on each of the blocks shown in Figure 1.1. In these sections we make reference to Altera's Quartus II integrated design tool. Most FPLD design tools provide blocks shown in Figure 1.1 in one or several environments. Quartus II provides all the necessary utilities under one environment, which makes it easy to learn and is typical of a complete environment.

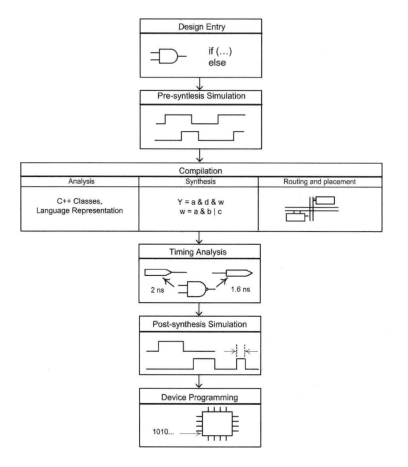

Figure 1.1 FPLD Design Flow

1.2 Design Entry

A design entry tool allows a designer to specify his or her design in textual and/or graphical form.   Generally, when specification of component interconnections is being done, a graphical entry tool suits best, while component behavior is best described by textual design entry methods. Whether to use a graphical or a textual design entry method also depends on the level of components being described and available parts. Usually, a design is specified by a mixture of graphical and textual representations, and design entry tools allow both schemes. Methods of design entry at various levels of hardware description are described in the following sub-sections.

1.2.1 Discrete Logic

A simple way of describing a design at the gate level is schematic entry using gate primitives. For this purpose, a schematic entry tool allows selection of gates and provides tools for wiring gates. The resulting circuit description can be used for simulation, synthesis and device programming.

Figure 1.2 shows a two-gate design in the schematic entry program of Quartus II. In this design, IO pins are used to mark and label inputs and outputs of the design.

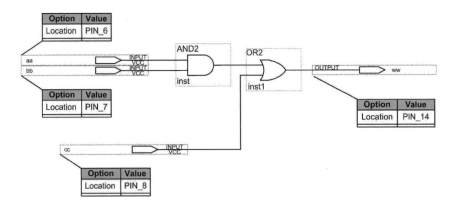

Figure 1.2 Discrete Logic Entry Tool

For simple designs and logic used for gluing together larger components (glue logic) this entry method is appropriate. However, for larger designs it is impossible to manually place all gates and specify their interconnections. For large gate level designs, basic components are built by use of gate-level primitives, and then these components are hierarchically used to complete the design.

1.2.2 Pre-Designed Components

After being involved in several designs, a hardware designer usually forms a library of hardware functions that the designer can use in his or her next designs. Designers usually test such components, document them and place them in a library for future use. Design team members working on different parts of a design, share such design libraries.

Figure 1.3 shows a component from a user library wired together with discrete gates. The mechanism for wiring library components is the same as that of primitive gates as discussed in relation with Figure 1.2.

User components can only be used as pre-defined library components if a symbol is made for them. In a design entry tool, a symbol editor program allows generation of a custom symbol for a design. Quartus II allows a custom symbol generation as well as automatic generation of a default symbol.

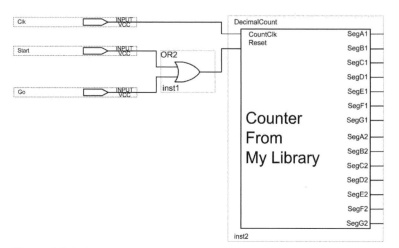

Figure 1.3 Using a Library Component in a Schematic Specification

1.2.3 Configurable Parts

In addition to allowing the use of primitives and pre-designed components, a schematic entry tool has a set of pre-defined library components of its own. Altera refers to these components as megafunctions or Mega-blocks. Mega-blocks can be designs consisting of as few as 2 or 3 gates to complete programmable processors.

To make these libraries useful to designers with a wide range of requirements, these blocks are made configurable. When a designer chooses a certain configurable part (megafunction), the design entry environment asks for the size of inputs, outputs, clocking scheme, and many other options specific to the component being configured.

1.2.4 Generic Configurable Functions

Some of the very common configurable parts are adders, ALUs, counters, stacks, queues and processors. When a designer selects a counter, the schematic entry program allows the user to specify, parameters like counter size, clocking, parallel-load, set and resetting, and carry out.

As an example, Figure 1.4 shows a configuration window of a counter megafunction. After the component is configured, it can be placed in the schematic editor of a design entry program and wired with other parts and components.

1.2.5 Configurable Memories

Megafunctions are for functions that are generic and have a wide range of applications. However, a designer may require functions that are hard to implement with discrete logic and at the same time are not generic enough to be able to use megafunctions for their implementation. In such cases, designers have the choice of using a ROM (for combinational circuits) or a RAM (for memory functions) for implementing their designs.

As an example, consider a design that reads keyboard codes and generates ASCII codes. For this, a large ROM can do the lookup of the ASCII code.

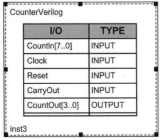

Figure 1.4 Configuring a Counter

User interface of our schematic entry program allows the use of ROM megafunctions. For this, the user specifies the number of rows, columns, and input or output clocking of the ROM. In addition, an initialization file is used for specifying ROM contents.

1.2.6 HDL Entry

With the increasing complexity of digital systems, the use of Hardware Description Languages (HDL) has become an essential mechanism for design entry.

CounterVerilog

I/O	TYPE
CountIn[7..0]	INPUT
Clock	INPUT
Reset	INPUT
CarryOut	INPUT
CountOut[3..0]	OUTPUT

inst3

Figure 1.5 HDL Interface Symbol

Tools for FPLD design, allow the use of VHDL and Verilog for design specification. One way of using an HDL description in a design is to take a complete description of a part, generate a symbol of it and use it like any other predefined component in the design. Alternatively, a schematic entry tool, such as that of Quartus II, allows definition of interface of an HDL part and uses its wiring mechanisms to wire this HDL part with other design components. Symbolic representation of this method is shown in Figure 1.5.

After defining the interface symbol, Quartus II allows generation of an HDL template that a designer can use to enter his or her HDL code. The template consists of the name of the component and its input and output ports.

1.3 Simulation

An important utility in any digital design environment is its simulation tool. There are two ways a design can be simulated. One is pre-synthesis simulation of an HDL description for functional and behavioral verification, and the other is post-synthesis simulation for detailed timing verification.

1.3.1 Pre-Synthesis Simulation

Before a design described in Verilog or VHDL is synthesized, its functionality must be verified. This verification is for discovering design errors, specification problems and incompatibility of parts used in a design.

Because high-level HDL designs are usually described at the level that specifies system registers and transfer of data between registers through busses, this level of system description is referred to as Register Transfer Level (RTL). Pre-synthesis simulation is also referred to as RT-level simulation.

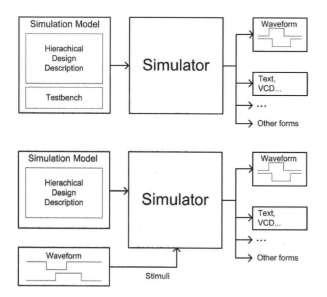

Figure 1.6 Test Data for Simulation, Using a Testbench, and Waveform Editor

At the RT level a design includes clock level timing but no gate and wire delays are included. Simulation at this level is accurate to the clock level. By performing RT level simulation, hazards, glitches, race conditions, setup and hold violations and other timing issues will not be detected. The advantage of this simulation is its speed compared with the gate level simulation.

Simulation of a design requires test data and HDL simulation environments provide various methods for application of this data to the design being tested. Test data can be generated graphically using waveform editors, or by use of HDL testbenches. Figure 1.6 shows two alternatives for defining test input data for a simulation engine. Outputs of simulators are in the form of waveforms (for visual inspection) and text for large designs for machine processing.

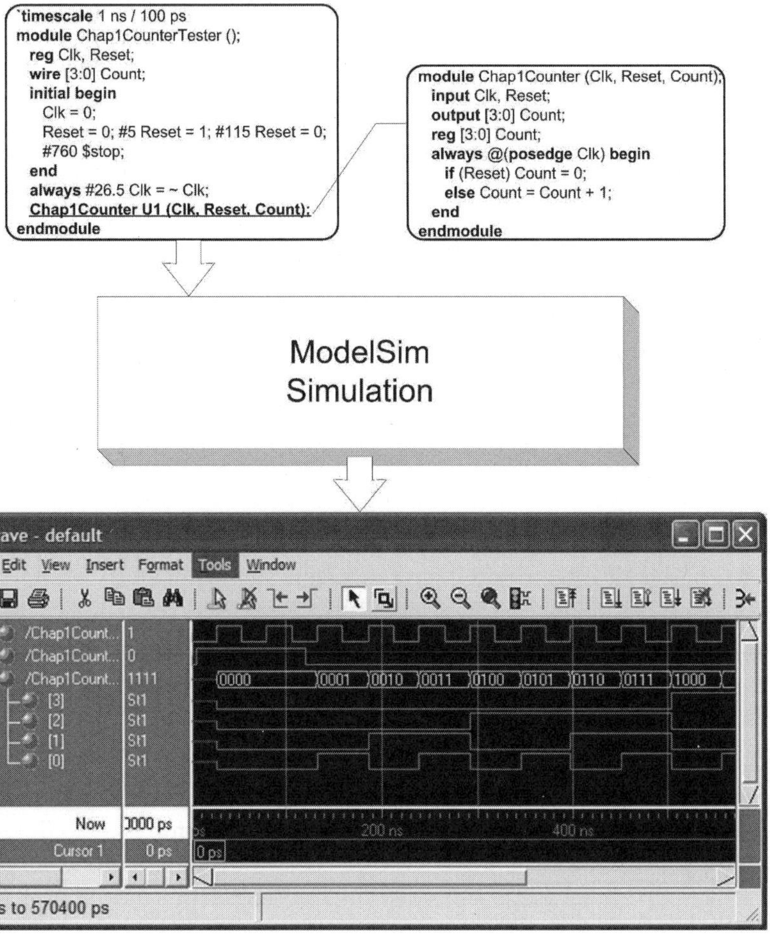
`timescale 1 ns / 100 ps
module Chap1CounterTester ();
reg Clk, Reset;
wire [3:0] Count;
initial begin
Clk = 0;
Reset = 0; #5 Reset = 1; #115 Reset = 0;
#760 $stop;
end
always #26.5 Clk = ~ Clk;
Chap1Counter U1 (Clk, Reset, Count);
endmodule
module Chap1Counter (Clk, Reset, Count);
input Clk, Reset;
output [3:0] Count;
reg [3:0] Count;
always @(posedge Clk) begin
if (Reset) Count = 0;
else Count = Count + 1;
end
endmodule
ModelSim
Simulation
wave - default
File Edit View Insert Format Tools Window
0000 0001 0010 0011 0100 0101 0110 0111 1000
Now
Cursor 1
0 ps
200 ns
400 ns
0 ps to 570400 ps

Figure 1.7 HDL Simulation with ModelSim

Simulation using an HDL testbench uses a testbench that instantiates the design under test, and as part of the code of the testbench it applies test data to the instantiated circuit. The complete testbench and the design instantiated within it are referred to as the simulation model. The simulation model that is shown in the upper part of Figure 1.6 is taken by the simulator engine, analyzed and processed and its results are generated for visual inspection and/or machine validation. Quartus II environment suggest use of ModelSim HDL simulator for VHDL and Verilog simulation. Figure 1.7 shows a Verilog code of a counter circuit, its testbench and its simulation results as simulated by ModelSim.

As shown in Figure 1.7, simulation results validate functionality of the counter circuit being tested. With every clock pulse the counter is incremented by 1. Note in the timing diagram shown that the counter output changes with the rising edge of the clock and no gate delays and propagation delays are shown here. The simulation shown in this figure shows good results no matter how fast the circuit clock frequency is.

Obviously, an actual hardware component behaves differently. Based on the timing and delays of the parts used, there will be a non-zero delay between the active edge of the clock and the counter output. Furthermore, if the clock frequency applied to an actual part is too fast for propagation of values within the gates and transistors of a design, the output of the design becomes unpredictable.

The simulation shown here is not provided with the details of timing of the hardware being simulated. Therefore, timing problems of the hardware that are due to gate delays cannot be detected. This is typical of a pre-synthesis or high-level behavioral simulation. What is being verified in Figure 1.7 is that our counter counts binary numbers. How fast the circuit works and what clock frequency it requires can only be verified after the design is synthesized.

1.3.2 Post-Synthesis Simulation

Timing issues, determination of a proper clock frequency and race and hazard considerations can only be checked by a post-synthesis simulation run after a design is synthesized. After synthesizing a design, details of gates used for the implementation of the design as well as wiring delays and load effects become evident. The simulation model used for post-synthesis simulation contains all such information.

The compilation phase of a design flow (Figure 1.1) generates a netlist of gates used along with timing files. Quartus II has an embedded simulator for post-synthesis simulation. This simulation is much slower than pre-synthesis simulation because it analyses the design at the gate level. Waveforms and simulation results show delay values between signal changes. If hazards occur, they appear as glitches in the simulation report of a post-synthesis simulation.

Figure 1.8 shows a Quartus II waveform editor screen for specification of inputs of our counter example, and shows the waveform generated by post-synthesis simulation in Quartus II.

As shown in this figure, there is a delay between the rising edge of the clock and the counter output. The figure shows 5.1 ns between the rising edge of the clock and the time that *count[1]* changes. As shown, the delay is slightly different for *count[0]*. Also depending on the count value, the delay value may vary.

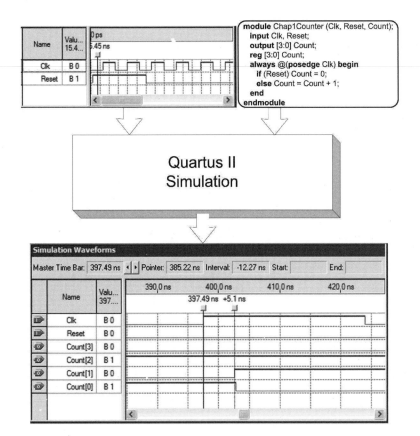

Figure 1.8 Post-Synthesis Simulation in Quartus II

Due to delays of wires and gates, it is possible that the behavior of a design as intended by the designer and its behavior after post-synthesis simulation are different. In this case, the designer must modify his or her design and try to avoid close timings and race situations.

1.3.3 Timing Analysis

As shown in Figure 1.1, as part of the compilation process, or in some tools after the compilation process, there is a timing analysis phase. This phase generates worst-case delays, clocking speed, delays from one gate to another, as well as required setup and hold times. Results of timing analysis appear in tables and / or graphs. Designers use this information to decide on their clocking speed and, in general, speed of their circuits.

Quartus II allows users to place timing specification and constraints on the implementation of their designs. This timing will be considered when the final layout of the design is being done on an FPLD chip. Due to the internal delays of programmable devices, such user constraints are not always satisfied.

1.4 Compilation

After a design is successfully entered and its pre-synthesis simulation results have been verified by the designer, it must be compiled to make it one step closer to an actual hardware on silicon.

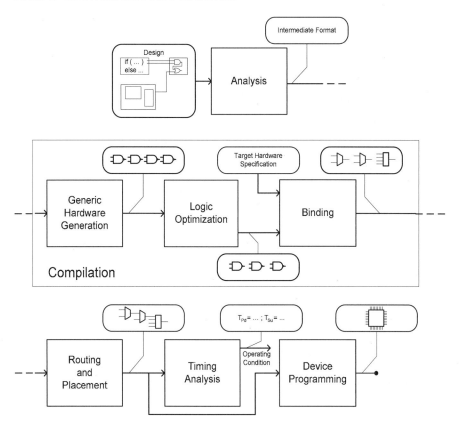

Figure 1.9 Compilation Process

The compilation process, translates various parts of a design that are described by various methods of data entry to an intermediate format (analysis phase), links all parts together, generates the corresponding logic (synthesis phase), places and routes parts on the target FPLD and generates timing details.

Figure 1.9 shows the compilation process and a graphical representation for each of the compilation phase outputs. As shown, the input of this phase is a hardware description that consists of HDL and schematic descriptions, and its output is a detailed hardware for programming an FPLD. Information about the specific FPLD to be programmed (target hardware specification in Figure 1.9) enter the compilation process at its binding stage.

1.4.1 Analysis

As discussed in Section 1.2, different parts of a design may be entered by various design entry methods. A complete design may consist of VHDL code, Verilog code, gates, or parts described in some propriety tool vendor's format. Before the complete design is turned into hardware, the design must be analyzed and a uniform format must be generated for all parts of the design.

In the analysis phase, HDL code syntax and semantics, use of proper interconnections between components, and appropriate use of pre-defined components will be checked.

1.4.2 Generic Hardware Generation

After a uniform presentation for all components of a design is obtained, the synthesis pass begins its operation by turning the design into a generic hardware format, such as a set of Boolean expressions or a netlist of basic gates.

1.4.3 Logic Optimization

The next phase of synthesis, after a design has been converted to a set of Boolean expressions, is the logic optimization phase. This phase is responsible for reducing expressions with constant input, removing redundant logic expressions, two level minimization, and multi-level minimization that includes logic sharing.

This is a very computationally intensive process, and some tools allow users to decide on the level of optimization. Output of this phase is in form of Boolean expressions, tabular logic representations, or primitive gate netlists.

1.4.4 Binding

After logic optimization, the synthesis process uses information from target hardware to decide exactly what FPLD internal logic elements and cells are needed for the realization of the circuit that is being designed. This process is called binding and its output is specific to the FPLD used. Some FPLDs use multiplexers and some use look-up tables. After binding is done, interconnection of multiplexers or contents of memories implementing look-up tables will be determined.

1.4.5 Routing and Placement

The routing and placement phase decides on the placement of the FPLD cells selected in the binding phase. Specific cells of the FPLD and wiring of inputs and outputs of these cells through wiring channels and switching areas are determined by the routing and placement phase.

The output of this phase is specific to the FPLD being used as target hardware and can be used for its programming. Altera FPLDs use *.sof* (SRAM Object File) and *.pof* (Programming Object File) formats for programming their FPLDs.

The timing analyzer that was discussed in Section 1.3.3 uses post routing and placement information.

1.5 Device Programming

Programmable devices have configuration elements that are programmed to make a device perform the functionality we program into it. Furthermore, there are various ways these configuration elements can be programmed. This section discusses configuration elements and device programming hardware alternatives.

1.5.1 Configuration Elements

Programmable devices incorporate three types of configuration elements: EEPROM, SRAM, and EPROM. EEPROMs and EPROMs are non volatile memories, and SRAM is volatile. All three types are reprogrammable.

EEPROM. The EEPROM cell is a transistor that is either ON or OFF depending on the threshold voltage. Unlike EPROM devices, however, EEPROM devices can be erased electrically. The EEPROM cell consists of a single, floating polysilicon gate structure that is used to change the threshold voltage of the transistor. The threshold voltage is changed when a tunneling mechanism traps an access of electrons on the floating gate. Once the electrons have been trapped on the floating gate, they present a negative shielding voltage and increase the threshold voltage of the transistor, making it impossible to turn the transistor on under normal operating voltages. This process allows the floating gate to act as an ON/OFF switch for the read transistor.

The EEPROM cell is erased by the tunneling mechanism. That is, electrons are removed from the floating gate, and the gate has a net positive charge that allows the EEPROM transistor to be turned on or off, depending on the voltage on the control gate.

SRAM. SRAM configuration elements are standard Static CMOS memory cells that consist of NMOS and PMOS transistors. Address lines enable writing data into cross-coupled gates of the memory through its input line, while the same address lines enable reading data from this static memory element through its output. Programming such devices is simply, writing into the memory and the memory element remains programmed for as long as the power is not removed from it.

EPROM. The EPROM transistor is a modified NMOS transistor in which the threshold voltage is easily switched between a low voltage and a high voltage. The different threshold voltages represent the EPROM cell in the ON and OFF states. The EPROM transistor has a floating polysilicon gate between the access gate (the regular transistor gate) and the substrate. The floating gate is electrically isolated from the substrate, on the one side, and the access gate, on the other side. EPROM transistors are programmed to a high-threshold voltage with hot electron injection. This electron injection causes some electrons to be trapped on the floating-gate electrode, creating a net negative voltage on the floating gate that opposes the electric filed created by the positive voltage on the access gate. The result is a substantial increase in the threshold voltage required to change the EPROM cell from a non-conducting to a conducting state by its access gate. The programmed EPROM cell behaves as a transistor that is turned off, and an erased cell works like a regular transistor operating by its access gate.

Programmed EPROM cells in the OFF state are erased by exposing the device to ultraviolet (UV) radiation. The excess electrons on the floating gate absorb radiant UV energy, experience a rise in energy level, overcome the oxide-silicon potential barrier, and finally migrate into substrate where they are neutralized.

1.5.2   Programming Hardware

Programmable device manufacturers offer a variety of hardware to program and configure their devices. For conventional device programming, in-system programming, and in-circuit reconfiguration, designers can choose from external programming hardware, external PC based devices, stand-alone programmers, and download cables. Devices are programmed by sending serial data generated after a design is successfully synthesized. The bit stream that is generated as configuration or programming data is sent to the programmable device in order to program it.

PC Based Programmers. A PC-based programmer is a hardware module that is used together with an appropriate adapter to program programmable devices. Such a component connects to a PC via the Universal Serial Bus (USB) or the serial port. Programming and functional test information is transmitted from the PC through the USB or serial port connection to the programmer.

Stand-Alone Programmers. A stand-alone programmer, together with the appropriate programming adapters, provides the hardware and software needed for programming EPROM- and EEPROM-based devices and configuring SRAM-based devices.

Download Cables. An inexpensive way of programming devices is to download serial configuration data via a serial download cable into the programmable device. Such cables interface to either a standard PC or UNIX workstation RS-232 port, USB port or the parallel port. Serial configuration or programming data is sent to the JTAG port of a configurable device via the download cable.

Configuration Devices. With SRAM-based devices, configuration data must be reloaded each time the system initializes, or when new configuration data is needed. Configuration devices store configuration data for SRAM-based programmable devices. Configuration devices are programmed by any of the programming hardware mentioned above and interface with the SRAM based device to transfer their non-volatile program into them. The configuration data going from a configuration device to an SRAM-based programmable device is clocked and serial.

1.6 Summary

This chapter gave an overview of mechanisms, tools, and processes used for taking a design from the design stage to an FPLD implementation. This overview contained information that will become clearer in the chapters that follow. We tried to make this information as generic as possible and not bound to a specific tool or environment. However, as a typical environment, specific references to the terminologies used by Quartus II were made.

2 Logic Design Concepts

This chapter gives a review of basic logic design concepts. The purpose is to highlight only those topics that are essential for design. Knowledge of the theoretical concepts, and much of the background concepts are assumed here. The chapter begins with a review of number systems, and basic logic gates. Combinational circuits and design of combinational circuits are discussed next. We will then focus on memory elements and sequential circuit design. Because of importance of state machines in RT level designs, special attention is given to these circuits in this chapter.

2.1   Number Systems

The transistor is the basic element of all digital electronic circuits. A transistor in a digital circuit behaves as an on-off switch. Because all elements are based on this on-off switch, they only take two distinct values. These values can be (ON, OFF), (TRUE, FALSE), (3V, 0V), or **(1, 0)**.

Because of this two-value system, all numbers in a computer are in base-2 or binary system. On the other hand, we use the decimal system in our every day life. To be able to understand what happens inside a digital system, we have to be able to convert between base-10 (Decimal) and base-2 (Binary) systems.

2.1.1   Binary Numbers

A decimal number has n digits and the weight of each digit is 10^i, where i is the position of digits counting from the right hand side and starting with 0. For example, 3256 is evaluated as:

$$(3256)_D = 6*10^0 +5*10^1 +2*10^2 +3*10^3 = 3256$$

A number in base-2 is evaluated similarly, except that the weights in decimal are 2^i instead of 10^i. For example 10110 is evaluated as:

$$(10110)_B = 0*2^0 +1*2^1 +1*2^2 +0*2^3 +1*2^4 = 22$$

By considering the weights in decimal and multiplying **bi**nary digits (bit) by their weights a binary number is converted to its equivalent decimal.

For conversion from decimal to binary, a decimal number is broken into necessary 2^i parts. Corresponding to the i values for which the decimal number has a 2^i part, there is a **1** in the equivalent binary number. For example $(325)_D$ can be broken into:

256 that is 2^8,
 64 that is 2^6,
  4   that is 2^2, and
  1   that is 2^0.

Therefore, the equivalent binary number has **1**s in positions 0, 2, 6, and 8, which makes the binary equivalent of $(325)_D$ to become $(101000101)_B$.

Methods described above for decimal to binary and binary to decimal also apply to fractional numbers. In this case the weight of digits on the right hand side of the decimal point are 10^{-1}, 10^{-2}, 10^{-3}, ... Similarly, the weights of binary digits on the right hand side of the binary point of a fractional binary number are 2^{-1}, 2^{-2}, 2^{-3}, ...

For example, $(1101.011)_B$ in binary becomes $(13.375)_D$ in decimal, and $(19.7)_D$ in decimal translates to $(10011.101)_B$. When converting from decimal to binary, for keeping the same precision as in the decimal number, a fractional decimal digit translates to 3 fractional binary digits.

2.1.2   Hexadecimal Numbers

A number in binary requires many bits for its representation. This makes, writing, documenting, or entering into a computer very error-prone. A more compact way of representing numbers, while keeping a close correspondence with binary numbers, is Hexadecimal representation.

Table 2.1 shows Hexadecimal digits and their equivalent Decimal and Binary representations. As shown, a base-16 digit translates to exactly 4 bits. Because of this, conversion from (to) a binary number to (from) its hex (hexadecimal) equivalent is a straight forward process. Therefore, we can use Hex numbers as a compact way of writing binary numbers. Several examples are shown below:

$(10011.101)_B = (13.A)_H$
$(11101100)_B = (EC)_H$

Table 2.1 Hexadecimal Digits

Hex	Decimal	Binary
0	0	0000
1	1	0001
2	2	0010
3	3	0011
4	4	0100
5	5	0101
6	6	0110
7	7	0111
8	8	1000
9	9	1001
A	10	1010
B	11	1011
C	12	1100
D	13	1101
E	14	1110
F	15	1111

2.2 Binary Arithmetic

In general, binary arithmetic is done much the same way as it is in the decimal system. In straight arithmetic, binary arithmetic is even simpler than decimal because it only involves 1s and 0s.

2.2.1 Signed Numbers

As we discussed earlier, everything inside a digital system is represented by 1s and 0s. This means that we have no way of representing plus (+) or minus (-) signs for signed numbers other than using 1s and 0s. Furthermore, unlike writing on paper that we can use as many digits as we like, representing numbers inside a digital system is limited by the width of busses, storage units, and lines. Because of these, a binary number in a digital system uses a fixed width, and the left most bit of the number is reserved for its sign.

A simple signed number system is sign and magnitude (S&M) in which a 0 in the left-most position of the number represents a positive and a 1 represents a negative number. For example +25 in 8-bit S&M system is **00011001** and -25 is **10011001**. Note here that enough 0's are put between the sign-bit and the magnitude of the number to complete 8 bits.

2.2.2 Binary Addition

As mentioned before, binary addition is very similar to decimal addition, and even easier. Adding two numbers starts from the right-hand side and with addition of every two bits a carry is generated. The carry is added to the addition of the next higher order bits. An example binary addition is shown below.

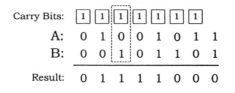

Addition is done in slices (bit positions) and with every add operation, there is a sum and a carry. The sum bit is the add result of the slice being added, and the carry is carried over to the next higher slice. The right-most bit result is the least-significant bit and is calculated first, and the sign-bit is calculated last.

2.2.3   Binary Subtraction

We can perform subtraction in binary using borrows from higher bits. This is similar to the way subtraction is done in decimal. However, this requires a different process from binary addition, which means that a different hardware is needed for its implementation.

2.2.4   Two's Complement System

As an alternative procedure for adding and subtracting, we can write numbers in the 2's complement number system and perform subtractions the same way we add. This signed number representation system is used to simplify signed number arithmetic.

Unlike the S&M system, in the 2's complement system just changing the sign-bit is not enough to change a positive number to a negative number or vice-versa. In this system, to change a positive (negative) number to a negative (positive) number, all bits must be complemented and a **1** must be added to it. For example -25 is calculated as shown below:

```
00011001   (=25)
11100110   (complementing all bits)
00000001   (adding a 1)
11100111   (-25)
```

When subtracting, instead of performing A-B, subtraction is done by A+(-B), in which (-B) is the two's complement of B. As an example consider subtraction of 25 from 93. First, 25 is turned into its two's complement negative representation that is **11100111** (as shown above). Then +93 that is **01011101** and -25 are added together as shown below:

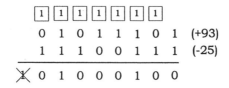

When adding a positive and negative number that results in a positive number, or adding two negative numbers that results in a negative number, a last carry (as in the above example) is generated that is ignored.

2.2.5 Overflow

In the two's complement arithmetic if adding two positive (negative) numbers with **0 (1)** sign bits results in a result that has a **1 (0)** in its sign-bit position, an overflow has occurred. This means that the result requires more room than is given to it. For example, the following addition is an over-flow case and the result is not valid.

```
              1  1  1  1  1  1
Sign Bits  1  0  1  1  1  0  1  0
           1  0  0  1  0  0  1  0
         ──────────────────────────
        X  0  0  0  1  1  0  0  0
```

In the above example the last bit beyond the sign-bit is dropped, as is done in 2's complement arithmetic. The final result of adding two negative numbers is a positive number that cannot be correct.

The case of overflow can be corrected by allocating more bits to the numbers involved in the two's complement arithmetic. A 2's complement number can be extended to occupy more bits by extending its sign-bit to the left. For example, 10111010 in 8-bit 2's complement system becomes 1111111110111010 in 16-bit 2's complement system. The overflow example shown above can be corrected if performed in 16-bit system as shown below:

```
   1  1  1  1  1  1  1  1  1  0  1  1  1  0  1  0
   1  1  1  1  1  1  1  1  1  0  0  1  0  0  1  0
 ─────────────────────────────────────────────────
X  1  1  1  1  1  1  1  1  0  1  0  0  1  1  0  0
```

In the above example, two negative numbers are added and a negative result is obtained, no over-flow occurs here.

2.3 Basic Gates

The transistor is the basic element for all digital logic components. However, for a design with several million transistors, designers cannot think at the transistor level. Therefore, transistors are put together into more abstract components, called gates, so that designers thinking at the high behavioral level can better relate to such abstract components. Later we will see that even gate structures are not abstract enough and designers need higher level means of specifying their designs. For this chapter, however, we concentrate on gates and gate-level designs.

2.3.1   Logic Value System

The (**0**, **1**) logic value system is a simple representation for voltage levels in a digital circuit.   However, this logic value system fails to represent many situations that are common in digital circuits.   For example if a line is connected to neither Gnd nor Vdd, it is neither **0** nor **1**. Or a line that is both driven by logic **0** and logic **1**, is neither a **0** nor a **1**.

A more complete system for representation of logic values is the four-value system, shown in Table 2.2.

Table 2.2 Four-Value Logic System

Value	Description
0	Forcing 0 or Pulled 0
1	Forcing 1 or Pulled 1
Z	Float or High Impedance
X	Uninitialized or Unknown

In logic simulations, a line that is not driven through pull-up or pull-down structures assumes **Z**. A line or a wire that is driven by both pull-up and pull-down structures appears as **X** in the simulation report.

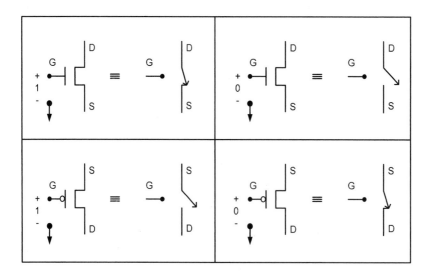

Figure 2.1 MOS Transistors

2.3.2 Transistors

The CMOS technology uses two types of transistors called NMOS and PMOS. These transistors act like on-off switches with the Gate input controlling connection (current flow) between Drain and Source terminals. As shown in Figure 2.1, an NMOS transistor conducts when logic **1** representing a high-voltage level drives its Gate. The conduction path allows current to flow between its Source and Drain terminals. Driving the Gate of an NMOS transistor with logic **0** (low voltage value) causes an open between Source and Drain terminals, which causes no current to flow through the transistor in either way.

As shown in Figure 2.1, opposite to the way an NMOS transistor works, the PMOS transistor conducts when its gate is driven by **0**, and is open when its gate is driven by logic **1** (or high voltage value).

2.3.3 CMOS Inverter

An inverter (also referred to as NOT gate) is a logic gate with an output that is the complement of its input. Transistor level structure of this gate, its logic symbol, its algebraic notations, and its truth table are shown in Figure 2.2.

In the transistor structure shown in this figure, if a is **0**, the upper transistor conducts and w becomes **1**. If a is **1**, there will be a conduction path from w to Gnd which makes it **0**. The table shown in Figure 2.2 is called the truth table of the inverter and lists all possible input values and their corresponding outputs. The inverter symbol is a bubble that can be placed on either side of a triangle representing a buffer.

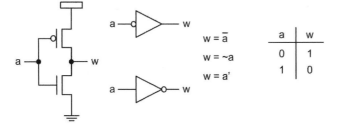

a	w
0	1
1	0

$w = \bar{a}$

$w = {\sim}a$

$w = a'$

Figure 2.2 CMOS Inverter (NOT gate)

2.3.4 CMOS NAND

A CMOS NAND gate uses two series NMOS transistors for pull-down, and two parallel PMOS transistors in its pull-up structure. Figure 2.3 shows structure and notations used for this gate.

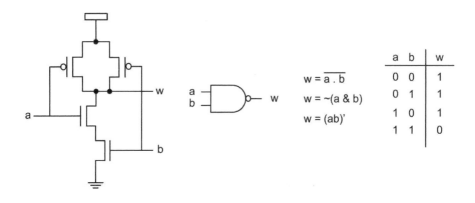

Figure 2.3 CMOS NAND

In the structure shown in Figure 2.3, if *a* and *b* are both **1**, there will be a conduction path from *w* to Gnd, making *w* **0**. Otherwise, the pull-up structure, instead of the pull-down structure, conducts that forces supply current to flow to *w*, making this output **1**.

2.3.5   CMOS NOR

A CMOS NOR gate uses two parallel NMOS transistors in its pull-down structure and two series PMOS transistors in its pull-up. Figure 2.4 shows structure and notations used for this gate. For the output of a NOR gate to become **1**, the pull-up structure must conduct. This means that both *a* and *b* must be **0**.

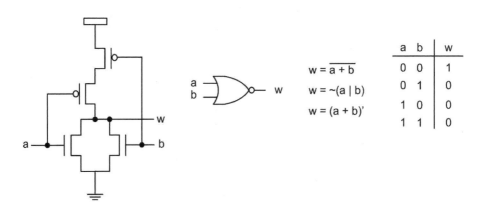

Figure 2.4 CMOS NOR

2.3.6 AND and OR gates

Figure 2.5 shows symbolic notations, algebraic forms and truth tables for AND and OR gates. These gates are realized using inverters on the outputs of NAND and NOR gates.

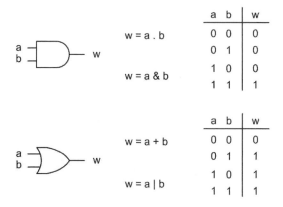

a	b	w
0	0	0
0	1	0
1	0	0
1	1	1

$w = a \cdot b$

$w = a \& b$

a	b	w
0	0	0
0	1	1
1	0	1
1	1	1

$w = a + b$

$w = a \mid b$

Figure 2.5 AND and OR gates

2.3.7 MUX and XOR gates

In addition to gates discussed above, several other logic structures become useful for realization of logic functions. One such gate or structure is the multiplexer that selects one of its inputs depending on the value of its select (s) input. Shown in Figure 2.6, the a input of the MUX appears on its output when s is **0**. HDL expression of the MUX and its truth table are shown in Figure 2.6. The right hand side of the equation shown reads as: *if (s is 1) then (b) else (a)*. This is a convenient conditional expression that is used in the C language and Verilog.

$w = s\,?\,b : a$

s	w
0	a
1	b

Figure 2.6 Multiplexer

The XOR gate (Exclusive-OR) that is shown in Figure 2.7 is similar to the OR gate except that its output is **1** when only one of its inputs is **1**.

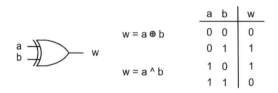

a	b	w
0	0	0
0	1	1
1	0	1
1	1	0

$w = a \oplus b$

$w = a \wedge b$

Figure 2.7 Exclusive-OR

2.3.8   Three-State Gates

All gates discussed so far generate a **1** or a **0** on their outputs depending on the values on their inputs. A three-state (also referred to as *tri-state*) buffer (or gate) has a data input (a) and a control input (c). Depending on c, it either passes a to its output (when c is **1**) or it floats the output (when c is **0**). As previously discussed, a float wire is represented by **Z**. Figure 2.8 shows a three-state buffer with true-value output and active-high control input. A truth-table and an algebraic representation are also shown for this structure.

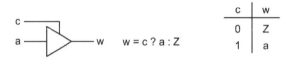

c	w
0	Z
1	a

$w = c\,?\,a:Z$

Figure 2.8 Three-State Gate

Other gate structures can be built by use of transistors arranged into complementary NMOS pull-down and PMOS pull-up structures. Furthermore, more complex functions can be built by use of gates discussed above.

2.4   Designing Combinational Circuits

Primitive gates discussed in the previous section form a set of structures with which any digital circuit can be designed. Methods of utilizing these parts for implementation of logic functions are discussed here.

2.4.1   Boolean Algebra

When a design is being done, a designer thinks of the functionality of the design and not the gate structure of it. To facilitate the use of logic gates and to make a correspondence between logic gates and design functions, Boolean algebra is used.

Boolean variables used in Boolean algebra take **1** or **0** values only. This makes Boolean algebraic rules different from the algebra that is based on decimal numbers.

Boolean algebra operations are AND, OR and NOT and their algebraic notations are ., + and ¬. The AND operator between two operands can be eliminated if no ambiguities occur. An over-bar also represents the NOT operator. Basic rules used for transformation of functions into gates are discussed below. These are Boolean algebra postulates and theorems.

1. $\bar{\bar{a}} = a$

2. $a + 0 = a$
   $a \cdot 1 = a$

3. $a + 1 = 1$
   $a \cdot 0 = 0$

4. $a + a = a$
   $a \cdot a = a$

5. $a + \bar{a} = 1$
   $a \cdot \bar{a} = 0$

6. $a + b = b + a$
   $a \cdot b = b \cdot a$

7. $a + (b + c) = (a + b) + c$
   $(a \cdot b) \cdot c = a \cdot (b \cdot c)$

8. $a + b \cdot c = (a + b) \cdot (a + c)$
   $a \cdot (b + c) = a \cdot b + a \cdot c$

9. $a + a \cdot b = a$
   $a \cdot (a + b) = a$

10. $a + \bar{a} \cdot b = a + b$
    $a \cdot (\bar{a} + b) = a \cdot b$

11. Duality: If E is true, changing AND (.) to OR (+), OR (+) to AND(.), 1 to 0, and 0 to 1 results in E_D that is also true.

12. DeMorgan's:
    $\overline{a \cdot b} = \bar{a} + \bar{b}$
    $\overline{a + b} = \bar{a} \cdot \bar{b}$

Once designers obtain functionality of their designs, they translate this functionality into a set of Boolean expressions. Using the above rules, this functionality can be manipulated, minimized, and put into a form that can be realized using gates described in Section 2.3.

As an example, consider the overflow situation that may arise in two's complement addition. Consider the sign bits of the operands and the result, $a7$, $b7$ and $s7$. Overflow (v) occurs if $a7$ is **1**, $b7$ is **1**, $s7$ is **0** or if $a7$ is **0**, $b7$ is

0, and $s7$ is **1**. This statement can be written as the following Boolean expression

$$v = a7 \cdot b7 \cdot \overline{s7} + \overline{a7} \cdot \overline{b7} \cdot s7$$

Applying Rule 12 (DeMorgan's Theorem), described above, v becomes:

$$v = \overline{\overline{a7 \cdot b7 \cdot \overline{s7}} \cdot \overline{\overline{a7} \cdot \overline{b7} \cdot s7}}$$

This expression is realized using NAND and NOT gates as shown in Figure 2.9.

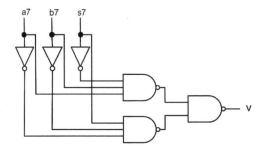

Figure 2.9 An Overflow Detector

2.4.2   Karnaugh Maps

Application of rules of Boolean algebra and expressing a hardware function with Boolean expressions is not always as easy as it is in the overflow example above. Karnaugh maps present a graphical method of representing Boolean functions. Karnaugh maps have close correspondence with tabular list of function output values, and at the same time present a visual method of applying Boolean algebra rules.

Figure 2.10 shows a 3-variable truth table and its corresponding karnaugh map (k-map). The truth table shows the listing of output values of a function in a list, and a k-map shows this information in a two-dimensional table.

A Boolean expression can be obtained for function f by reading rows of its truth table. As shown, function f is **1** for four combinations of a, b, and c. In Row #3, f is **1** if a is **0**, b is **1** and c is **1**. This means that the complement of a ANDed with b and ANDed with c make f become **1**. Therefore if $\overline{a} \cdot b \cdot c$ is **1** f becomes **1**. This term is called a product term and since it contains all variables of function f it is also called a minterm of this function. Corresponding to every row of f in which function f is **1** there is a minterm. Function f is **1** if any of its minterms are true. Therefore function f can be written by ORing its four minterms, as shown below:

$$f = \overline{a} \cdot b \cdot c + a \cdot \overline{b} \cdot c + a \cdot b \cdot c + a \cdot b \cdot \overline{c}$$

This form of representing a function is called sum of products, and since the product terms are all minterms, this representation is the Standard Sum Of Products (SSOP).

As shown in Figure 2.10, the same expression could be written by reading the karnaugh map shown. For this, a product term corresponds to every box of the Karnaugh map that contains a **1**. However, the k-map has certain properties that we can use to come up with a more reduced form of sum of products.

Row	a	b	c	f
0	0	0	0	0
1	0	0	1	0
2	0	1	0	0
3	0	1	1	1
4	1	0	0	0
5	1	0	1	1
6	1	1	0	1
7	1	1	1	1

b c \ a	0	1
00	0	0
01	0	1
11	1	1
10	0	1

f

Figure 2.10 A 3-Variable K-map

For discussion of Karnaugh map properties, we define Boolean and physical k-map adjacency as follows:

> *Boolean Adjacency:* Two product terms are adjacent if they consist of the same Boolean variables and only one variable appears in its true-form in one and complement in another (v in one, $\bar{v}$ in another).

> *Physical Adjacency:* Two k-map boxes are adjacent if they are horizontally or vertically next to each other.

Numbering k-map rows and columns are arranged such that input combinations corresponding to adjacent boxes in the map are only different in one variable. This means that two *Physical Adjacent* boxes are also *Boolean Adjacent.* The main idea in the k-map is that two minterms that are different in only one variable can be combined to form one product term that does not include the variable that is different in the two minterms.

In the k-map of Figure 2.10, $\bar{a} \cdot b \cdot c$ and $a \cdot b \cdot c$ that are Boolean adjacent can be combined into one product term as shown below:

$$\bar{a} \cdot b \cdot c + a \cdot b \cdot c = b \cdot c$$

In the resulting product term, variable a that appears as a in one product term and $\bar{a}$ in another is dropped. Because of adjacency in the k-maps, the same can be resulted without having to perform the above Boolean manipulations. Figure 2.11 shows minimization of function f using k-map grouping of terms.

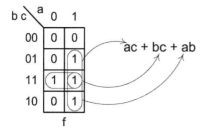

Figure 2.11 Minimizing Function f

This figure shows that instead of writing the ORing of minterms $\bar{a} \cdot b \cdot c$ and $a \cdot b \cdot c$ and then using Boolean algebra to reduce it to $b \cdot c$, we can directly write $b \cdot c$ by reading the k-map. Since the two 1's corresponding to these product terms are physically adjacent on the k-map, they are also Boolean adjacent. Therefore, the product term that corresponds to these two adjacent 1's is one that includes all the variables except the variable that appears as its true and its complement in the two adjacent k-map boxes (variable a).

The following Boolean manipulations correspond to the mappings shown in Figure 2.11:

$$f = \bar{a} \cdot b \cdot c + a \cdot \bar{b} \cdot c + a \cdot b \cdot c + a \cdot b \cdot \bar{c}$$
$$f = \bar{a} \cdot b \cdot c + a \cdot b \cdot c + a \cdot \bar{b} \cdot c + a \cdot b \cdot c + a \cdot b \cdot \bar{c} + a \cdot b \cdot c$$
$$f = (b \cdot c) \cdot (\bar{a} + a) + (a \cdot c) \cdot (b + \bar{b}) + (a \cdot b) \cdot (c + \bar{c})$$
$$f = (b \cdot c) \cdot (1) + (a \cdot c) \cdot (1) + (a \cdot b) \cdot (1)$$
$$f = b \cdot c + a \cdot c + a \cdot b$$

As shown above, the term $a \cdot b \cdot c$ is repeated 3 times. This is according to Boolean algebra Rule 4 of Section 2.4.1 that states ORing an expression with itself is the same as the original expression. In the k-map, application of this rule is implied by using the k-map box with a 1 that corresponds to $abc=111$ in as many mappings as we need (here in 3 mappings). For another example, we use a 4-variable map.

A four-variable function, its k-map, its minimal Boolean realization, and its gate level implementation are shown in Figure 2.12. To make a correspondence between Boolean adjacency and k-map physical adjacency, we visualize a k-map as a spherical map in which, in the back of the sphere, the sides of the map and its four corners are adjacent.

With this interpretation, the four corners of the k-map of Figure 2.12 form two product terms that are themselves adjacent. The complete mapping of the four corners of the map results in only one product term. By use of Boolean algebra rules, Figure 2.13 shows justification for combining the four corners of the k-map into one product term. In this diagram *Position* indicates North West, North East, South West and South East of the map.

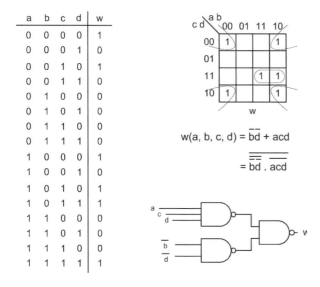

a	b	c	d	w
0	0	0	0	1
0	0	0	1	0
0	0	1	0	1
0	0	1	1	0
0	1	0	0	0
0	1	0	1	0
0	1	1	0	0
0	1	1	1	0
1	0	0	0	1
1	0	0	1	0
1	0	1	0	1
1	0	1	1	1
1	1	0	0	0
1	1	0	1	0
1	1	1	0	0
1	1	1	1	1

$$w(a, b, c, d) = \overline{b}\,\overline{d} + acd$$

$$= \overline{\overline{\overline{b}\,\overline{d}} \cdot \overline{acd}}$$

Figure 2.12 Minimizing a 4-variable Function

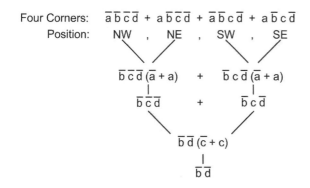

Four Corners: $\overline{a}\,\overline{b}\,\overline{c}\,\overline{d} + a\,\overline{b}\,\overline{c}\,\overline{d} + \overline{a}\,b\,c\,\overline{d} + a\,\overline{b}\,c\,\overline{d}$

Position: NW , NE , SW , SE

$$\overline{b}\,\overline{c}\,\overline{d}\,(\overline{a} + a) \quad + \quad \overline{b}\,c\,\overline{d}\,(\overline{a} + a)$$

$$\overline{b}\,\overline{c}\,\overline{d} \qquad + \qquad \overline{b}\,c\,\overline{d}$$

$$\overline{b}\,\overline{d}\,(\overline{c} + c)$$

$$\overline{b}\,\overline{d}$$

Figure 2.13 Combining Four Corners of a 4-variable Map

For implementation of this function another product term is needed to cover minterms $a \cdot b \cdot c \cdot d$ and $a \cdot \overline{b} \cdot c \cdot d$. Because of adjacency of these two terms, variable b drops in the resulting product term. Figure 2.12 shows the minimal realization of function w. After a minimal SOP is obtained, it is converted to an expression using NAND operations by application of DeMorgan's theorem.

2.4.3   Don't Care Values

In some designs, certain input values never occur or if they do occur, their outputs are not important. For example, consider a design that takes a one-digit BCD number (Binary Coded Decimal) as input and generates a **1** output when the input is divisible by 3. The 4-bit input includes combinations ranging from 0 to 15 binary. However, **1010** through **1111** combinations are not valid BCD numbers and are not expected to appear on the circuit inputs.

a	b	c	d	w
0	0	0	0	1
0	0	0	1	0
0	0	1	0	0
0	0	1	1	1
0	1	0	0	0
0	1	0	1	0
0	1	1	0	1
0	1	1	1	0
1	0	0	0	0
1	0	0	1	1
1	0	1	0	-
1	0	1	1	-
1	1	0	0	-
1	1	0	1	-
1	1	1	0	-
1	1	1	1	-

$$w\ (a, b, c, d) = \bar{a}.\bar{b}.\bar{c}.\bar{d} + \bar{b}.c.d + b.c.\bar{d} + a.d$$

Figure 2.14 Using Don't Care Values

When we are designing this circuit with a k-map, we have to decide what to do with the k-map boxes that correspond to the invalid BCD numbers. If we fill them with all **1**s, we will end up mapping unnecessary **1**s. However, if we fill them with all **0**s, mapping function minterms may become too limited. The alternative is to leave them as undecided or (Don't Care) values and let the mapping decide what values these invalid cases take.

We use a dash (-), or d or X for showing a Don't Care k-map value. When mapping for a minimal realization, we only use the Don't Care values if they help us form larger maps. This way, those mapped Don't Care values are used as **1**s and the rest are **0**s.

The solution to the problem stated above is shown in Figure 2.14. Note here that of the 6 Don't Care values 4 are used for forming larger maps and 2 are not mapped.

2.4.4   Iterative Hardware

Boolean minimization of functions by use of Boolean rules or, indirectly, by use of k-maps is only practical for small functions. Partitioning based on regularity of a structure, or based on independent functionalities, help in breaking a circuit into smaller manageable circuits.

For example, consider a 4-bit comparator that generates a **1** when its 4-bit *A* input is greater than its 4-bit *B* input (Figure 2.15). One way of doing this circuit is to try to come up with its minimal realization by doing an 8-variable k-map. Obviously, this is not practical.

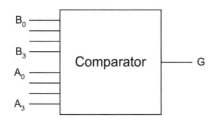

Figure 2.15 A 4-bit Comparator

Alternatively, we can design the comparator by first comparing the most significant bits of its two inputs and working our way into the least significant bit.

The G output becomes **1** if A_3 is greater than B_3. Logically, this means that the $A_3 \cdot \overline{B_3}$ product term forms an AND gate that is an input for an OR gate that generates G. Next, we compare A_2 and B_2 only if the decision for putting a **1** on G cannot be made by A_3 and B_3. This means that the decision based on A_2 and B_2 can only be made if A_3 and B_3 are equal. Therefore the $A_2 \cdot \overline{B_2}$ product term can only cause the G output to become **1** when A_3 and B_3 are equal $(A_3 \oplus B_3 = 1)$. Repeating this logic for all bits of the two inputs from bits 3 down to 0, we will cover all logics that cause G to become **1** when we reach A_0 and B_0. Figure 2.16 shows the resulting hardware for our 4-bit comparator. This hardware has a repeating part, and can easily be extended for larger magnitude comparators.

As another example of iterative hardware, consider the design of an 8-bit adder. Adding two 8-bit numbers is shown in Figure 2.17. As shown in this figure, adding is done bit-by-bit starting from the right-hand side. The process of adding repeats for every bit position. This process uses a carry-in from its previous position ($i\text{-}1$), adds it to A_i and B_i, and generates S_i as well as a carry-out for the next position. Therefore, hardware for the 8-bit adder uses eight repetitions of a one-bit adder that is called a Full-Adder (FA).

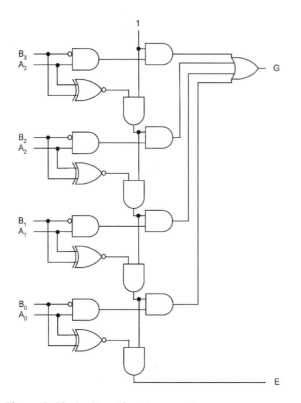

Figure 2.16 An Iterative Comparator

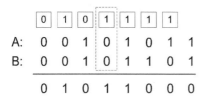

Figure 2.17 Adding Two 8-bit Numbers

The FA hardware has 3 inputs (Carry-in (c_i), bit i of A (a_i), and bit i of B (b_i)) and two outputs (Carry-out (c_o) and bit i of sum (s_i)). Figure 2.18 shows the design of an FA using k-maps. Also shown in this figure is an 8-bit adder using eight full-adders. This adder design is referred to as "ripple-carry" since the carry ripples from one FA to another.

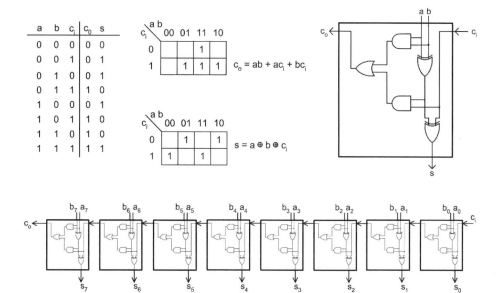

Figure 2.18 An 8-bit Ripple Carry Adder

Hardware components like the comparator and the adder described above are iterative, cascadable, extendable, and in many cases configurable.   In designing digital systems it is important to have a library of such packages available.   Instead of designing from scratch, a digital designer uses these packages and configures them to meet his or her design requirements.

Discrete logic gates used to match inputs and outputs of various packages are referred to as "Glue Logic".

2.4.5   Multiplexers and Decoders

Other packages that become useful in many high level designs include multiplexers and decoders.

A multiplexer is like an n-position switch that selects one of its n inputs to appear on the output. A multiplexer with n inputs is called an *n-to-1 Mux*. The number of bits of the inputs (b) determines the size of the multiplexer.   A multiplexer with n data inputs requires $s=log_2(n)$ number of select lines to select one of the n inputs; i.e., $2^s=n$.

For example, a multiplexer that selects one of its four ($n=4$) 8-bit ($b=8$) inputs is called an 8-bit 4-to-1 Mux. This multiplexer needs 2 select lines ($s=2$). Schematic diagram of this multiplexer is shown in Figure 2.19. This circuit can be built using an array of AND-OR gates or three-state gates wired to implement a wired-OR logic.   Figure 2.20 shows the gate level design of a 1-bit 4-to-1 Mux.

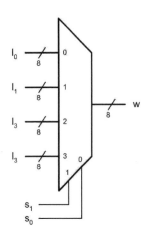

Figure 2.19 An 8-bit 4-to-1 Mux

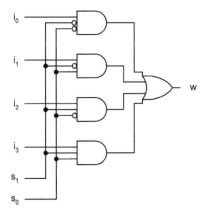

Figure 2.20 A 4-to-1 Mux

Multiplexers are used for data selection, bussing, parallel-to-serial conversion, and for implementation of arbitrary logical functions. A 1-bit 2-to-1 Mux can be wired to implement NOT, AND, and OR gates. Together with a NOT, a 2-to-1 Mux can be used for implementation of most primitive gates. Because of this property, many FPGA cells contain multiplexers for implementing logic functions.

Another part that is often used in high level designs is a decoder. Generally, a combinational circuit that takes a certain code as input and generates a different code is referred to as a decoder. For example, a circuit that takes as input a 4-bit BCD (Binary Coded Decimal) and generates outputs for display on a Seven Segment Display (SSD) is called a BCD to SSD decoder.

A more accurate definition is that a decoder has as many outputs as it has combinations of inputs. For every combination of values on its inputs a certain output of the decoder becomes active. For example a 2-to-4 binary decoder has 2 inputs forming four combinations. Its four outputs become active for input combinations, **00**, **01**, **10**, and **11**. The gate level design of this decoder is shown in Figure 2.21.

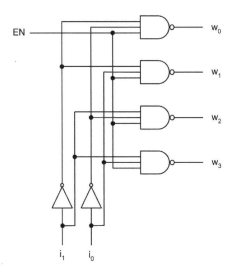

Figure 2.21 A 2-to-4 Decoder

The selected output in Figure 2.21 becomes **0** and all others are **1**. The circuit also has an enable input, *EN*. For the decoder to become operational, this input must be **1**. The enable input is useful for cascading decoders.

2.4.6 Activity Levels

Activity levels for input and output ports of digital circuits refer to the way that these ports function. For example an active-low output (like the decoder described above) is **1** when not active and it becomes **0** when active. An active-low enable input of a circuit makes the circuit operational when it is **0**. When such an input is **1**, circuit outputs become inactive. The *EN* input of the decoder described above is an active-high enable input.

A NAND gate can be looked at as an AND gate with an active-low output and active-high inputs. A NAND gate can also be looked at as an OR gate with active-low inputs and active-high output (see Figure 2.22). The following Boolean expressions justify these views of a NAND gate:

$$\overline{a \cdot b} = \neg(a \cdot b)$$
$$\overline{a \cdot b} = (\overline{a} + \overline{b})$$

Using correct polarities and notations with correct activity-level markings, make circuit diagrams more readable. For example in Figure 2.22 the two circuits with w output are equivalent. The one on the left requires writing Boolean expressions to understand its functionality, but the function of the one on the right can easily be understood by inspection.

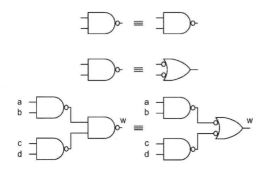

Figure 2.22 NAND Gate Activity Levels

2.4.7   Enable / Disable Inputs

Many digital logic packages, like multiplexers and decoders come with enable (*EN*) and/or output-enable (*OE*) inputs. When an input is referred to as *EN*, it means that if this input is not active, all circuit outputs are inactive. On the other hand, an *OE* input is for three-state control of the output. In a circuit with an *OE* input, if *OE* is active, the outputs of the circuit are as defined by the function of the circuit. However when *OE* is inactive, all circuit outputs become high-impedance or float (**Z** value).

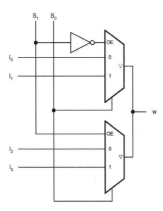

Figure 2.23 Wiring Circuits with OE Control Inputs

Circuits with three-state outputs require an *OE* input. Outputs of such circuits can be wired to form wired-OR logic. Figure 2.23 shows two 2-to-1 multiplexers with three-state outputs that are wired to form a 4-to-1 multiplexer.

If the multiplexers of Figure 2.23 had *EN* inputs instead of *OE* inputs, forming the final output of the circuit, *w*, would require an OR gate.

2.4.8   A High-Level Design

In the first part of this chapter we showed that instead of using transistors in a design, we wire them to form upper-level structures (primitive gates) with easier functionalities that digital designers can relate to. In the second part, we discussed the use of gates in still higher level structures such as adders, comparators, decoders and multiplexers. With these higher-level structures, designers will be able to think at a more functional level and not have to get involved in putting thousands of gates together for a simple design.

This level of design is called RT (Register Transfer) level. In today's designs, designers think at this level and most design tools work at this level. Most design libraries include configurable RTL components for designers to use.

As a simple RT level design, consider an 8-bit Absolute-Value calculator. The circuit takes a positive or negative 2's complement input and generates the absolute value of its input on its 8-bit output. The circuit diagram using RT level packages is shown in Figure 2.24.

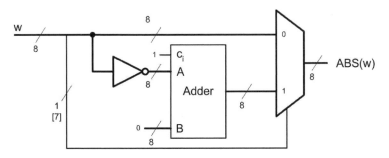

Figure 2.24 An Absolute Value Circuit

The circuit uses an array of eight NOT gates to form the complement of the input. Using the adder shown a **1** is added to this complement to generate the two's complement of the input. The multiplexer on the output uses the sign-bit of the input to select the input if it is positive or to select the 2's complement of the input if the input is a negative number.

2.5   Storage Elements

Circuits discussed so far in this chapter were combinational circuits that did not retain a history of events on their inputs. To be able to design circuits that can make decisions based on past events, we need to have circuits with memory that can remember some of what has happened on their inputs. This section discusses the use of memory elements that help us achieve this.

The past history of a memory circuit participates in determination of its present output values. Therefore, outputs of these circuits are not only a function of their inputs, but also a function of their past history. This history enters the logic structure of a memory circuit by way of feedbacks from its outputs back to its inputs. The more lines that are fed back means that the circuit remembers more of its past.

2.5.1 The Basic Latch

The circuit shown in Figure 2.25 is the basic latch. We will show that this circuit has some memory. The circuit has one feedback line from its y output back to its input. One feedback line that can take **0** or **1** binary values means that the circuit remembers only two things from its past.

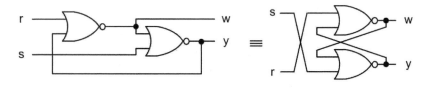

Figure 2.25 The Basic latch (Two Equivalent Circuits)

Applying the waveform shown in Figure 2.26 to the inputs of the latch of Figure 2.25 shows that a pulse on s sets the w output to **1** and a pulse on r sets it to **0**.

Note from of the timing diagram of Figure 2.26 that at time a when s and r are both **0**, w is **0**, and at time b when the same exact values appear on the circuit inputs the output is **1**. This reveals that the output depends not only on the present input, and that the circuit is remembering something from its past history.

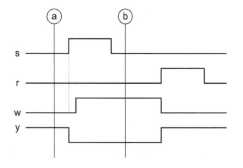

Figure 2.26 Setting and Resetting the Basic Latch

An interpretation of the behavior of this circuit is that a complete positive pulse on the input s causes w to set and a complete pulse on r causes it to reset. Because of this behavior, the circuit of Figure 2.25 is called an SR-Latch.

This structure is the basic element for most static memory structures. Alternative structures that implement this memory behavior use NAND gates or inverters and pass-transistors.

2.5.2   Clocked D Latch

The memory behavior of the SR-latch does not have a close correspondence to the way we think about storing data or saving information. The structure shown in Figure 2.27 improves this behavior. In this structure, when *clock* is **1** a **1** on D causes s to become **1** which causes Q to set to **1**, and a **0** on D causes r to become **1** to reset Q.

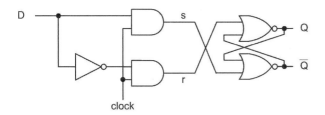

clock

Figure 2.27   A Clocked D-Latch

This structure behaves as follows: when *clock* becomes **1**, the value of D will be stored until the next time that *clock* becomes **1**. At all times this value appears on Q and its complement on $\overline{Q}$.

This behavior that at a given time, determined by the *clock*, a value is stored until the next time we decide to store a new value, corresponds more to the way we think about memories. The circuit of Figure 2.27 is called a clocked D-latch and is used in applications for storing data, buffering data, and temporary storage of data. For storing multiple bits of data, multiple latches with a common clock can be used. Figure 2.28 shows a quad latch using a symbolic representation of a latch.

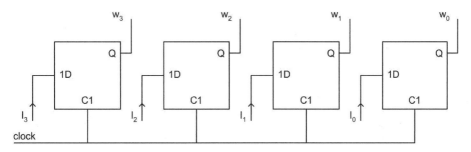

Figure 2.28   Quad Latch

In Figure 2.27 when *clock* is 1, data on D pass through the latch and reach Q and changes on D directly affect Q. Because of this, this structure is called a transparent latch. The symbolic notation of latch shown in Figure 2.28 indicates dependence of D on *clock*. This shows that control signal *1* that is the *clock* signal controls the D input.

2.5.3   Flip-Flops

The latch as discussed above is a good storage element, but because of its transparency, it cannot be used in feedback circuits. Take for example a situation that the output Q of the latch goes through a combinational circuit and feeds back to its own inputs (see Figure 2.29). Because latches are transparent, the feedback path stays open while the clock signal is active. This will result in an unpiedictable latch output due to the uncontrolled number of times that data feeds back through the logic block. In some cases the output oscillates while the clock is active.

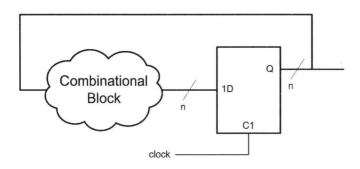

Figure 2.29 Latch Feedback Causes Unpredictable Results

To overcome the above mentioned problem, a structure without transparency must be used. Conceptually this is like the use of double doors for building entrances. At any one time only one door is open to keep the air-conditioned air inside the building.

For our case, we use two latches with inverting clocks as shown in Figure 2.30. When *clock* is **0**, the first latch stops data on D from propagating to the output. When *clock* becomes **1**, data is allowed to propagate only as far as the output of the first latch (M). While this is happening the second latch stops data on M from propagating any further. As soon as *clock* becomes **0**, D and M are disconnected and data stored in M propagates to Q. The latch on the left is called master, and the one on the right is the slave. This structure is called a master-slave D-flip-flop. At all times, input and output of this structure are isolated.

Other forms of flip-flops that have this isolation feature use a single edge of the clock to accept the input data and affect the output. Such structures are called edge-trigger flip-flops. Figure 2.31 shows a rising- and a falling-edge D-flip-flop. The triangle indicates edge triggering and the bubble on the clock input of the circuit on the right indicates negative (falling) edge triggering.

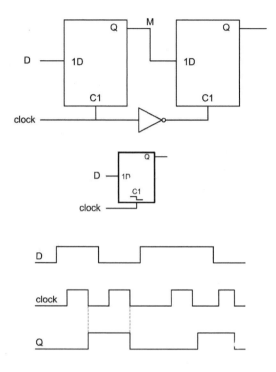

Figure 2.30 A D-Flip-Flop, Its Symbolic Notation and Waveform

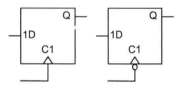

Figure 2.31 Edge Trigger Flip-flops

2.5.4   Flip-Flop Control

The initial value of a flip-flop output depends on its internal gate delays, and in most cases is unpredictable. To force an initial state into a flip-flop, set and reset control inputs should be used. Other control inputs for flip-flops are clock-enabling and three-state output control signals.

A *Set* or *Preset* control input forces a flip-flop into its **1** state, and a *Reset* or *Clear* input forces it to **0**. We refer to these signals as flip-flop initialization inputs. Such control inputs can act independent of the clock, or act like the D-input with the specified edge of the clock. In the former case, the initialization inputs must be put into the internal logic of a flip-flop and are called asynchronous control inputs. In the case that a control signal only affects the flip-flop when the flip-flop is clocked, it is called a synchronous control input. Synchronous control inputs can be added to a flip-flop by adding external logic. Figure 2.32 shows four flip-flops with various forms of synchronous and asynchronous controls. To indicate clock dependency in a flip-flop with a synchronous control, the clock identifier (number *1* on the right hand side of letter *C*) is used on the left hand side of the control signal name.

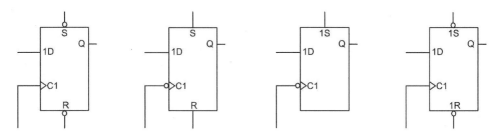

Figure 2.32 Flip-flops with Synchronous and Asynchronous Control

Another control input for flip-flops is a clock enabling input. When enabled, the flip-flop accepts its input when a clock pulse arrives, and when disabled, clocking the flip-flop does not change its state.

Figure 2.33 Clock Enabling

Two implementations for clock enabling are shown in Figure 2.33. The one on the left, circulates data back into the flip-flop when it is disabled (*EN* = 0). When enabled, the external data on the *D* input is clocked into the flip-flop. The structure shown on the right, uses an AND gate to actually gate the clock and stop the flip-flop from accepting data on its *D* input. This is called clock gating and because of its critical timing issues, care must be taken when using this implementation.

Some flip-flops come with three-state outputs. In this case, a three-state buffer on the output is controlled by an *OE* (Output Enable) control input. Hardware implementation of this feature and its symbolic notation are shown in Figure 2.34. The use of a triangle on the output side of the symbolic notation is useful, but is not always used.

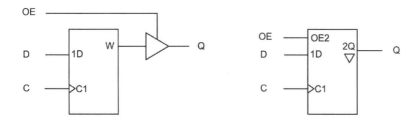

Figure 2.34 Three-State Control

2.5.5 Registers

The structure formed by a group of flip-flops with a common clock signal and common control signals is called a register. As with flip-flops, registers come in different configurations in terms of their enabling, initialization and output control signals. Figure 2.35 shows an 8-bit register with an active-low three-state output control and a synchronous active low reset. A register is also said to a group of latches.

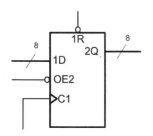

Figure 2.35 An 8-bit Register

2.6 Sequential Circuit Design

This section discusses design of circuits that have memory; such circuits are also called sequential circuits. We will first discuss the design of sequential circuits using discrete parts (gates and flip-flops) and then focus our attention on sequential packages. This approach is similar to what was done in Section 2.4 for combinational circuits.

2.6.1 Finite State Machines

A sequential circuit is a digital system that has memory and decisions it makes for a given input depend on what it has memorized. These circuits have local (inside flip-flops) or global feedbacks and the number of feedbacks determine how much of its past history it remembers.

The number of states of a sequential circuit is determined by its memory. A circuit with n memory bits has 2^n possible states. Signals or variables representing these states (*n* of them) are called state variables. Because sequential circuits have a finite number of states, they are also called finite-state machines, or FSM.

All sequential circuits - from a single latch to a network of high performance computers - can be regarded as an FSM. These machines can be modeled as a combinational circuit with feedback. If the feedback path includes an array of flip-flops with a clock for controlling the timing of data feeding back, the circuit becomes a synchronous sequential circuit. Figure 2.36 shows the Huffman model of synchronous sequential circuits. This model divides such a circuit into a combinational part and a register part.

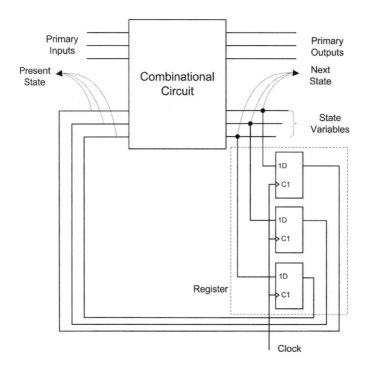

Figure 2.36 Huffman Model of a Sequential Circuit

The clock shown is the synchronization signal. Outputs that are fed back to the inputs are state variables. The inputs of the flip-flops become the present state of the machine after the circuit clock ticks. The circuit decides on its outputs and its next state based on its inputs and its present state.

2.6.2   Designing State Machines

To show the design process for FSMs, we use a simple design with one input and one output. The circuit searches on its input for a sequence of **1**s and **0**s.

This circuit is called a sequence detector, and the procedure used in its design applies to the design of very large FSMs.

Problem Description. A sequence detector with one input, x and one output, w, is to be designed. The circuit searches on its x input for a sequence of **1011**. If in four consecutive clocks the sequence is detected, then its output becomes **1** for exactly one clock period. The circuit continuously performs this search and it allows overlapping sequences. For example, a sequence of **1011011** causes two positive pulses on the output. Figure 2.37 shows a timing diagram example of this search.

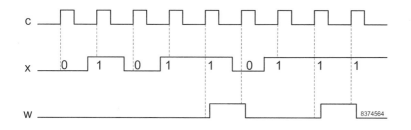

Figure 2.37 Searching for 1011

State Diagram. The above problem description is complete, but does not formally describe the machine. To design this sequence detector, a state diagram which has representations for all states of the machine must be used. A state diagram is like a flowchart and it completely describes our state machine for values that occur on its input. Input events are only considered if they are synchronized with the clock. Figure 2.38 shows the state diagram of our **1011** detector.

As shown in this state diagram, each state has a name (A through E) and a corresponding output value (w is **1** in E and **0** in the other states). There are edges out of each state for all possible values of circuit inputs.

Since we only have one input, two edges, one for $x=0$ and one for $x=1$ are shown for each state. Since the machine is to detect **1011**, this sequence always ends in state E, no matter what state we start from. In each state, if the input value that takes the machine one state closer to the output is not received (e.g., receiving a **0** in state D), the machine goes to the state that saves the most number of bits of the correct sequence. For example a **0** in state D takes the machine to state C that has a **0** output, since state D is the state that **101** has been detected and a **0** on x makes the remembered received bits **1010**. Of these remembered bits only the last **10** can be used towards a correct sequence, and therefore the machine goes to state C that remembers this sequence.

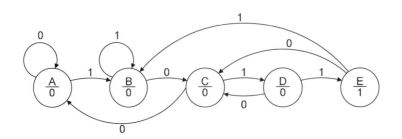

Figure 2.38   State Diagram for the 1011 Detector

State Table. Design of digital circuits requires data and behavior of the circuit that is being designed to be represented in a tabular form. This enables us to form truth tables and/or k-maps from this behavioral description. Therefore, the next step in design of our sequence detector is to form a table from the state diagram of Figure 2.38.

	x		
State	0	1	
A	A	B	0
B	C	B	0
C	A	D	0
D	C	E	0
E	C	B	1
	State⁺	w	

Figure 2.39   State Table of the 1011 Detector

Figure 2.39 shows the state table that corresponds to this state diagram. The first column shows the present states of the machine, *State*. Table entries are the next states of the machine for *x* values **0** and **1**. The table also shows the output of the circuit for various states of the machine. State *E* goes to state *C* for *x* of **0** and to state *B* if *x* is **1**. The value of the *w* output in this state is **1**.

State Assignment. The state table of Figure 2.39 takes us one step closer to the hardware implementation of our sequence detector, because the information is represented in a tabular form instead of the graphical form of Figure 2.38. However, hardware implementation requires all variables in a circuit description to be in binary. Obviously, in our state table, state names are not in binary.

For this binary representation, we assign a unique binary pattern (binary number) to each of the states of our state table. This step of the work is called "state assignment". Because we have five states, we need five unique binary numbers, which means that we need three bits for giving our states unique bit patterns. Figure 2.40 shows the state assignment that we have decided to use for this design.

State	y_2	y_1	y_0
A	0	0	0
B	0	0	1
C	0	1	0
D	0	1	1
E	1	0	0

Figure 2.40 State Assignment

Specific bit patterns given to the states of a state machine are not important. Binary values assigned to each state become values for y_2, y_1, and y_0. These variables are state variables of our machine.

Transition Table. Now that we have binary values for the states of our state machine, state names in the state table of Figure 2.39 must be replaced with their corresponding binary values. This will result in a tabular representation of our circuit in which all values are binary. This table is called a transition table and is shown in Figure 2.41.

	y_2	y_1	y_0	X 0	1	
A:	0	0	0	0 0 0	0 0 1	0
B:	0	0	1	0 1 0	0 0 1	0
C:	0	1	0	0 0 0	0 1 1	0
D:	0	1	1	0 1 0	1 0 0	0
E:	1	0	0	0 1 0	0 0 1	1
				$y_2{}^+ \ y_1{}^+ \ y_0{}^+$		w

Figure 2.41 Transition Table for the 1011 Detector

A transition table shows the present values of state variables (y_2, y_1 and y_0) and their next values ($y_2{}^+$, $y_1{}^+$, $y_0{}^+$). Next state values are those that are assigned to the state variables after the circuit clock ticks once. Since only five of eight possible states are used, three combinations of state-variable value are unused. Therefore, next state and output values for these table entries are don't care values.

Excitation Tables.   So far in the design of the **1011** detector, we have concentrated on the design of the complete circuit including its combinational and register parts, as defined by the Huffman model of Figure 2.36. We have been discussing present and next state values, which obviously imply a sequential circuit.

In the next step of the design, we separate the combinational and the register parts of the design. The register part is simply an array of flip-flops with a common clock signal. The combinational part is where present values of flip-flops (their outputs) are used as input to generate flip-flop input values that will become their next state values.

Because a D-type flip-flop takes the value on its D input and transfers it into its output after the edge of clock, what we want to become its next state is the same as what we put on its D input. This means that the required D input values generated by the combinational part of a sequential circuit are no different than their next state values ($Q^+ = D$). Therefore, tables for values of D_2, D_1 and D_0 in our **1011** sequence detector are the same as those for y_2^+, y_1^+, and y_0^+. Flip-flop input tables are called excitation tables that are shown in Figure 2.42 for our design.

			x		
y_2	y_1	y_0	0	1	
0	0	0	0 0 0	0 0 1	0
0	0	1	0 1 0	0 0 1	0
0	1	0	0 0 0	0 1 1	0
0	1	1	0 1 0	1 0 0	0
1	0	0	0 1 0	0 0 1	1
1	0	1	- - -	- - -	-
1	1	0	- - -	- - -	-
1	1	1	- - -	- - -	-
			$D_2\ D_1\ D_0$		w

Figure 2.42 Flip-flop Excitation Tables

Implementing the Combinational Part.   Now that we have separated the combinational and register parts of our design, the next step is to complete the design of the combinational part. This part is completely described by the table of Figure 2.42. This table includes values for D2, D1 and D0 in terms of x, y2, y1, and y0, as well as values for w in terms of y2, y1 and y0. Karnaugh maps shown in Figure 2.43 are extracted from the table of Figure 2.42.

Figure 2.43 also shows Boolean expressions for the D-inputs of y_2, y_1 and y_0 flip-flops. The four-input (x, y_2, y_1, and y_0), four-output (w, D_2, D_1, and D_0) combinational circuit is fully defined by Boolean expressions of Figure 2.43.

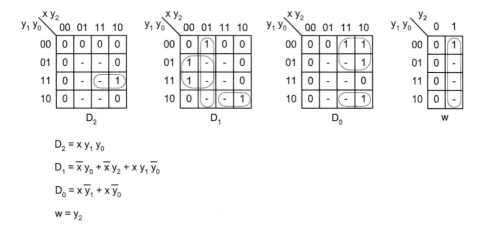

$$D_2 = x\, y_1\, y_0$$

$$D_1 = \overline{x}\, y_0 + \overline{x}\, y_2 + x\, y_1\, \overline{y_0}$$

$$D_0 = x\, \overline{y_1} + x\, \overline{y_0}$$

$$w = y_2$$

Figure 2.43 Implementing Combinational Part

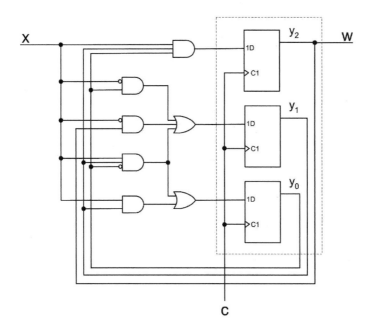

Figure 2.44 Logic Block Diagram of the 1011 Detector

Complete Implementation. The design of the **1011** sequence detector will be completed by wiring the gate-level realization of the combinational part with the flip-flops of the register part. This realization is shown in Figure 2.44.

Implementation of the **1011** detector is according to the Huffman model of Figure 2.36. The box on the left is the combinational part, and the one on the right is the register part. State variables of this circuit are y_2, y_1 and y_0 that are fed back from the outputs of the combinational part back into its inputs through the register part. The clocking mechanism and initialization of the circuit only affect the register part. For asynchronous initialization of the circuit, flip-flops with asynchronous set and/or reset inputs should be used. For synchronous initialization, AND gates on the D inputs should be used for resetting and OR gates for setting the flip-flops.

2.6.3 Mealy and Moore Machines

The design presented in the previous section produces an output that is fully synchronous with the circuit clock. In its state diagram, since the output is specified in the states of the machine, while in a given state, the output is fixed. This can also be seen in the circuit block diagram of Figure 2.44 in which the logic of the w output only uses the state variables, and does not involve x. This state machine is called a Moore machine. A more relaxed timing can be realized by use of a different machine that is referred to as a Mealy machine.

Figure 2.45 shows the Mealy state diagram of the **1011** detector. As shown, the output values in each state are specified on the edges out of the states, along with input values. This means that while in a given state, the value on x decides the value of the output. For example, in state D, if x is **0**, w is **0** and if x is **1**, w is **1**.

With this dependency, changes on x propagate to the output even if they are not accompanied by the clock. The implementation of a Mealy machine is similar to that of a Moore machine, except that the output k-map involves the inputs as well as the state variables. A sequence detector that is implemented with a Mealy machine usually requires one state less than the Moore machine that detects the same sequence. If implemented as a Mealy machine, our detector requires four states, two state variables, and three 3-variable Karnaugh maps for the two state variables and the output.

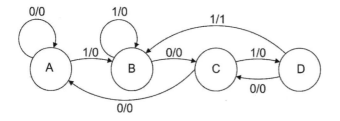

Figure 2.45 Mealy State Diagram

2.6.4 One-Hot Realization

Instead of going through steps discussed in Section 2.6.2 for gate-level implementation of a state machine, a more direct realization can be obtained by using one flip-flop per state of the machine. Since in a state diagram only one state is active at any one time, only one of the corresponding flip-flops becomes active. This method of state assignment is called *one-hot* assignment. This implementation uses more flip-flops than the binary state assignment discussed in Section 2.6.2, but uses fewer logic gates for activation of the flip-flops.

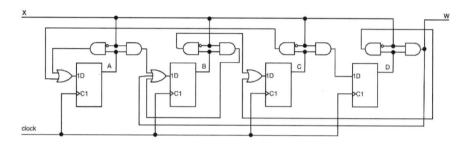

Figure 2.46 One-hot Implementation

One-hot implementation of the Mealy machine of Figure 2.45 is shown in Figure 2.46. Output of the AND gates on the outputs of the flip-flops correspond to the edges that come out of the states of the state diagram. These AND gates are conditioned by $x=0$ or $x=1$. The four flip-flops used yield 2^4 possible states. Of these 16 states only four are used (**1000, 0100, 0010** and **0001**). Initialization of a one-hot machine should be done such that it is put into one of its valid states. Starting the machine in **0000** is wrong because it will never get out of this state.

Some of the advantages of one-hot machines are their ease of design, regularity of their structure, and testability.

2.6.5 Sequential Packages

As there are commonly used combinational packages, like adders, decoders and multiplexers, there are commonly used sequential packages like registers, counters and shifters. An RT level designer first partitions his or her design into such predefined components, and will only resort to designing with discrete components when there are no packages that meet the design requirements.

Counters. Counters are used in many RT level designs. A counter is a sequential circuit that counts a certain sequence in ascending or descending order. An n-bit binary up-counter counts n-bit numbers in the ascending order.

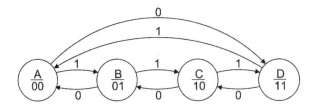

Figure 2.47 State Diagram of a 2-bit Counter

As an example we will show the design of a 2-bit up-down counter. With each clock pulse, when UD is **1** it counts up and when UD is **0** it counts down. In the count-up mode the next count after **11** is **00**, and in the count-down mode the next count after **00** is **11**.

The state diagram for this counter is shown in Figure 2.47. Counter count outputs are shown in each state. This is a Moore state machine and the procedure discussed earlier in this chapter can be used for its design. However, because of the simple sequencing of counter circuits, many of the steps discussed in Section 2.6.2 can be skipped and we can go directly from the description of the counter to its transition tables. Furthermore, if we decide to use D-type flip-flops for our counter, excitation tables, or even D-input k-maps, can be written based on the count sequence. Figure 2.48 shows k-maps generated directly from the up and down sequences of the counter of Figure 2.47.

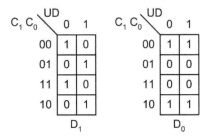

Figure 2.48 Excitation K-maps for a 2-Bit Up-Down Counter

In the right columns of the k-maps when $UD=1$, values for $D1$ and $D0$ are set to take C_1 and C_0 through the **00, 01, 10, 11**, ... sequence. In the left columns of the k-maps, D_1 and D_0 values make the counter count the **11, 10, 01, 00**, ... sequence. Circuit shown in Figure 2.49 performs the basic up- and down-countings for our 2-bit counter.

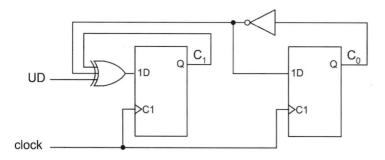

Figure 2.49 A Two-Bit Up-Down Counter

In addition to the basic counting implemented by the circuit of Figure 2.49, other features in a counter include, resetting, parallel loading, enabling, carry-in and carry-out. Resetting a counter is like resetting registers. Asynchronous resetting forces the counter into its initial state and acts independent of the clock. Synchronous resetting loads the initial state of the counter through the D-inputs of counter flip-flops, which obviously requires the proper clocking of the register.

To start counting from a given state, the counter is put into parallel-load mode and the designated start state is loaded into the flip-flops of the counter. In this mode the counter acts just like a register. Inputs of flip-flops of a counter with parallel load feature must be available outside of the counter package.

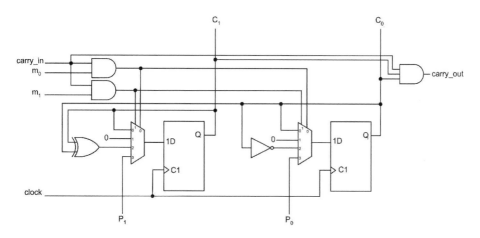

Figure 2.50 Two-Bit Up-Counter with Added Features

An enable input for a counter makes it count only when this input is active. This signal controls clocking of data into the individual flip-flops of the counter.

Some counters have carry-in and carry-out input and output signals that are used for cascading several of them. Carry-out output of a modulo-n

counter becomes **1** when the counter reaches its maximum count. The carry-in
input of a counter (if it exists) acts just like an enable input except that it also
enables the carry-out of the counter. Figure 2.50 shows a two-bit up-counter
with added features of synchronous reset, enable, parallel load, carry-in and
carry-out.

The m_1 and m_0 inputs of the counter shown in Figure 2.50 are its mode
inputs. These inputs control data that are clocked into the flip-flops. If mode is
0 (m_1, m_0 = 0, 0), the counter is disabled. In mode 1 the counter resets to 0, in
mode 2 the counter counts up. Mode 3 is for parallel load; in this mode P_1 and
P_0 are loaded into the counter. The counter only counts if *carry_in* is **1**,
otherwise it is disabled. When *carry_in* is **1** and counter reaches **11**, the
carry_out becomes **1**. Cascading counters can be done by connecting *carry_out*
of one to the *carry_in* of another.

Shifters. Shift registers are registers with the property that data shifts right or
left with the edge of the clock. Shift registers are used for serial data collection,
serial to parallel, and parallel to serial converters.

Figure 2.51 shows a 4-bit right shifter. With every edge of the clock data in
the register moves one place to the right. Data on S_i (serial-in) starts moving
into the register and data in the register moves out bit-by-bit from S_0 (serial-
out).

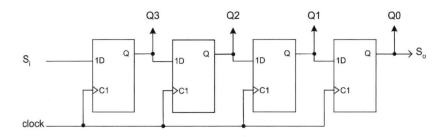

Figure 2.51 A 4-bit Shift Register

Shift-registers can be easily cascaded by connecting S_0 of one to the S_i of
another. Other functionalities included in these packages are left-shift, parallel
load, enable, and reset. These features can be included in much the same way
as in counters (Figure 2.50). Shift-registers with three-state output control use
three-state gates on their output, like what is done in registers (Figure 2.34 and
Figure 2.35).

2.7 Memories

In their simplest form, memories are two-dimensional arrays of flip-flops, or
one-dimensional arrays of registers. Flip-flops in a row of memory share read
and write controls, and memory rows share input and output lines.

The number of flip-flops in a row of memory is its word-length, m. Memory words are arranged so that each word can individually be read or written into. Memory access is limited to its words. A memory of 2^n m-bit words has n address lines for addressing and enabling read and write operations into its words. The address space of such a memory is 2^n words. Input and output busses of such a memory have m bits. The block diagram of a clocked memory with a r/w (read/write) control signal is shown in Figure 2.52. The CE input shown is the Chip Enable input, which must be active for the memory to be read or written into.

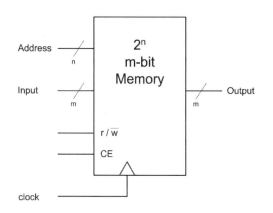

Figure 2.52 A 2^n m-bit Memory

Because accessing words in a memory can be done independent of their location in the memory array and by simply addressing them, memories are also called RAM or Random Access Memory. RAM structures come in various forms, SRAM (Static RAM), DRAM (Dynamic RAM), Pseudo-Static RAM, and many other forms that depend on their technology as well as their hardware structures.

2.7.1 Static RAM Structure

Figure 2.53 shows an SRAM that has an address space of 4, and word length of 3. The address bus for this structure is a 2-bit bus ($2^2 = 4$), and its input and output are 3-bit busses. A 2-to-4 decoder is used for decoding the address lines and giving access to the words of the memory. An external Chip-Enable disables all read and write operations when it is **0**.

The logic of the decoder shown in Figure 2.53 may be distributed inside the memory array. Other blocks in the memory shown are a read-write logic block and an IO block.

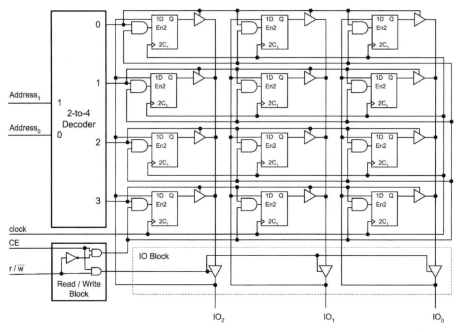

Figure 2.53 SRAM Structure

2.7.2   Bidirectional IO

The memory shown in Figure 2.53 has bidirectional *inout* lines used both as input and output. In the input mode, *IO* lines feed D-flip-flop inputs. In the output mode, three-state gates in the IO buffer block take the output of the addressed memory word and put it on the *IO* of the memory.

Bidirectional *inout* lines are useful for cascading memory chips and for reducing pin count of memory packages.

2.8   Summary

This chapter presented an overview of basic logic design. The focus was mostly on the design techniques and not on their theoretical background. We covered combinational and sequential circuits at the gate and RT levels. At the combinational gate-level, we discussed Karnaugh maps, but mainly concentrated on the use of iterative hardware and packages. In the sequential part, state machines were treated at the gate level; we also discussed sequential packages such as counters and shift-registers. The use of these packages facilitates RT level designs and use of HDLs in design.

3 Verilog for Simulation and Synthesis

This chapter presents Verilog from the point of view of a designer wanting to describe a design, perform pre-synthesis simulation, and synthesize his or her design for programming an FPGA or generating a layout. Many of the complex Verilog constructs related to timing and fine modeling features of this language will not be covered here. The chapter first describes Verilog with emphasis on design using simple examples. We will cover the basics, just enough to describe our examples. In a later section after a general familiarity with the language is gained, more complex features of the Verilog language with emphasis on testbench development will be described.

3.1 Design with Verilog

Verilog syntax and language constructs are designed to facilitate description of hardware components for simulation and synthesis. In addition, Verilog can be used to describe testbenches, specify test data and monitor circuit responses. Figure 3.1 shows a simulation model that consists of a design and its testbench in Verilog. Simulation output is generated in form of a waveform for visual inspection or data files for machine readability.

After a design passes basic functional validations, it must be synthesized into a netlist of components of a target library. Constructs used for verification of a design, or timing checks and timing specifications are not synthesizable. A Verilog design that is to be synthesized must use language constructs that have a clear hardware correspondence. Figure 3.2 shows a block diagram specifying the synthesis process.

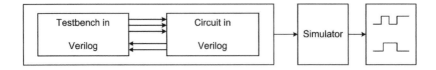

Figure 3.1 Simulation in Verilog

The output of synthesis is a netlist of components of the target library. Often synthesis tools have an option to generate this netlist in Verilog. In this case, the same testbench prepared for pre-synthesis simulation can be used with the netlist generated by the synthesis tool.

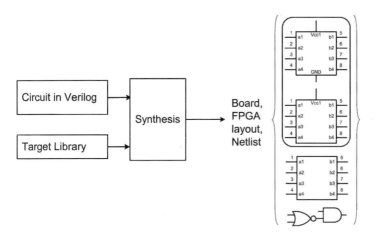

Figure 3.2 Synthesis

3.1.1   Modules

The entity used in Verilog for description of hardware components is a **module**. A module can describe a hardware component as simple as a transistor or a network of complex digital systems. As shown in Figure 3.3, modules begin with the **module** keyword and end with **endmodule**.

```
module
...
...
endmodule
```

Figure 3.3 Module

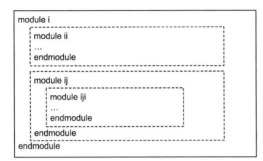

Figure 3.4 Module Hierarchy

A design may be described in a hierarchy of other modules. The top-level module is the complete design, and modules lower in the hierarchy are the design's components. Module instantiation is the construct used for bringing a lower level module into a higher level one. Figure 3.4 shows a hierarchy of several nested modules.

As shown in Figure 3.5, in addition to the **module** keyword, a module header also includes the module name and list of its ports. Following the module header, its ports and internal signals and variables are declared. Specification of the operation of a module follows module declarations.

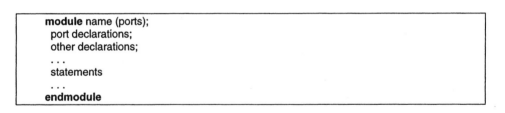

Figure 3.5 Module Outline

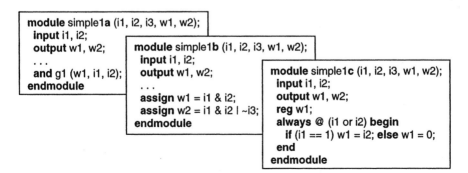

Figure 3.6 Module Definition Alternatives

Operation of a module can be described at the gate level, using Boolean expressions, at the behavioral level, or a mixture of various levels of abstraction. Figure 3.6 shows three ways operation of a module may be described. Module *simple1a* in Figure 3.6 uses Verilog's gate primitives, *simple1b* uses concurrent statements, and *simple1c* uses a procedural statement.

The subsections that follow describe details of module ports and description styles. In the examples in this chapter Verilog keywords and reserved words are shown in bold. Verilog is case sensitive. It allows letters, numbers and special character "_" to be used for names. Names are used for modules, parameters, ports, variables, and instance of gates and modules.

For readability of graphics, we use the symbol shown in Figure 3.7 for representing a Verilog module. Inputs are shown as hollow boxes, and outputs as solid ones. The name of the module appears inside the module box on its upper side.

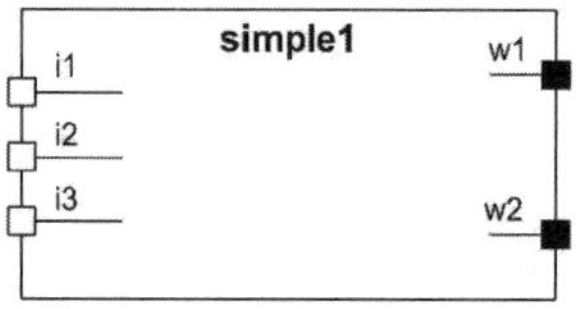

Figure 3.7 Module Notation

3.1.2   Module Ports

Following the name of a module is a set of parenthesis with a list of module ports. This list includes inputs, outputs and bidirectional input lines. Ports may be listed in any order. This ordering can only become significant when a module is instantiated, and does not affect the way its operation is described. Top-level modules used for testbenches have no ports.

```
module acircuit (a, b, c, av, bv, cv, w, wv);
  input a, b;
  output w;
  inout c;
  input [7:0] av, bv;
  output [7:0] wv;
  inout [7:0] cv;
  . . .
endmodule
```

Figure 3.8 Module Ports

Following module header, ports of a module are declared. In this part, size and direction of each port listed in the module header are specified. A port may be **input**, **output** or **inout**. The latter type is used for bidirectional input/output lines. Size of vectored ports of a module is also declared in the module port declaration part. Size and indexing of a port is specified after its type name within square brackets. Figure 3.8 shows an example circuit with scalar, vectored, input, output and inout ports. Ports named *a*, and *b* are one-bit inputs. Ports *av* and *bv* are 8-bit inputs of *acircuit*. The set of square brackets that follow the **input** keyword applies to all ports that follow it. Port *w* of *acircuit* is declared as a 1-bit output, and *wv* is an 8-bit bi-directional port of this module.

```
module bcircuit (a, b, av, bv, w, wv);
  input a, b;
  output w;
  input [7:0] av, bv;
  output [7:0] wv;
  wire d;
  wire [7:0] dv;
  reg e;
  reg [7:0] ev;
  . . .
endmodule
```

Figure 3.9 Wire and Variable Declaration

In addition to port declarations, a module declarative part may also include wire and variable declarations that are to be used inside the module. Wires (that are called **net** in Verilog) are declared by their types, **wire**, **wand** or **wor**; and variables are declared as **reg**. Wires are used for interconnections and have properties of actual signals in a hardware component. Variables are used for behavioral descriptions and are very much like variables in software languages. Figure 3.9 shows several wire and variable declarations.

```
module vcircuit (av, bv, cv, wv);
  input [7:0] av, bv, cv;
  output [7:0] wv;
  wire [7 :0] iv, jv;
  assign iv = av & cv;
  assign jv = av | cv;
  assign wv = iv ^ jv;
endmodule
```

Figure 3.10 Using Wires

Wires represent simple interconnection wires, busses, and simple gate or complex logical expression outputs. When wires are used on the left hand sides of **assign** statements, they represent outputs of logical structures. Wires can be used in scalar or vector form. Figure 3.10 shows several examples of wires used on the right and left hand sides of **assign** statements.

In the vector form, inputs, outputs, wires and variables may be used as a complete vector, part of a vector, or a bit of the vector. The latter two are referred to as part-select and bit-select.

3.1.3   Logic Value System

Verilog uses a 4-value logic value system. Values in this system are **0**, **1**, **Z**, and **X**. Value **0** is for logical **0** which in most cases represent a path to ground (Gnd). Value **1** is logical **1** and it represents a path to supply (Vdd). Value **Z** is for float, and **X** is used for un-initialized, undefined, un-driven, unknown, and value conflicts. Values **Z** and **X** are used for wired-logic, busses, initialization values, tri-state structures, and switch-level logic.

For more logic precision, Verilog uses strengths values as well as logic values. Our dealing with Verilog is for design and synthesis, and these issues will not be discussed here.

3.2   Combinational Circuits

A combinational circuit can be represented by its gate level structure, its Boolean functionality, or description of its behavior.   At the gate level, interconnection of its gates are shown; at the functional level, Boolean expressions representing its outputs are written; and at the behavioral level a software-like procedural description represents its functionality. This section shows these three levels of abstraction for describing combinational circuits.

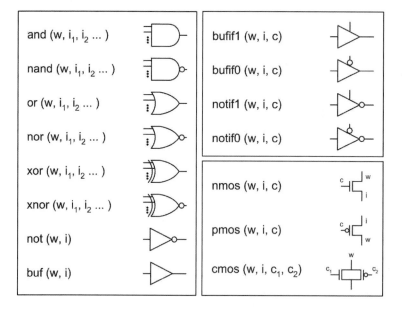

Figure 3.11 Basic Primitives

3.2.1 Gate Level Combinational Circuits

Verilog provides primitive gates and transistors. Some of the more important Verilog primitives and their logical representations are shown in Figure 3.11. In this figure w is used for gate outputs, i for inputs and c for control inputs.

Basic logic gates are **and, nand, or, nor, xor, xnor**. These gates can be used with one output and any number of inputs. The other two structures shown, are **not** and **buf**. These gates can be used with one input and any number of outputs.

Another group of primitives shown in this figure are three-state (tri-state is also used to refer to these structures) gates. Gates shown have w for their outputs, i for data inputs, and c for their control inputs. These primitives are **bufif1, notif1, bufif0,** and **notif0**. When control c for such gates is active (**1** for first and third, and **0** for the others), the data input, i, or its complement appears on the output of the gate. When control input of a gate is not active, its output becomes high-impedance, or **Z**.

Also shown in Figure 3.11 are NMOS, PMOS and CMOS structures. These are switches that are used in switch level description of gates, complex gates, and busses. The **nmos (pmos)** primitive is a simple switch with an active high (low) control input. The **cmos** switch is usually used with two complementary control inputs. These switches behave like the three-state gates. They are different in their output voltage levels and drive strengths. These parameters are modeled by wire strengths and are not discussed in this book.

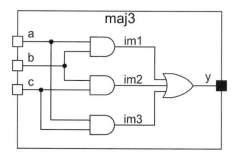

Figure 3.12 A Majority Circuit

Majority Example. We use the majority circuit of Figure 3.12 to illustrate how primitive gates are used in a design. The description shown in Figure 3.13 corresponds to this circuit. The module description has inputs and outputs according to the schematic of Figure 3.12.

Line 1 of the code shown is the **timescale** directive. This defines all time units in the description and their precision. For our example, $1ns/100Ps$ means that all numbers in the code that represent a time value are in nanoseconds and they can have up to one fractional digit (100 Ps).

The statement that begins in Line 6 and ends in Line 9 instantiates three **and** primitives. The construct that follows the primitive name specifies rise and fall delays for the instantiated primitive ($t_{plh}=2$, $t_{phl}=4$). This part is optional and if eliminated, 0 values are assumed for rise and fall delays. Line 7 shows inputs and outputs of one of the three instances of the **and** primitive. The output is *im1* and inputs are module input ports *a* and *b*. The port list on Line 7 must be followed by a comma if other instances of the same primitive are to follow, otherwise a semicolon should be used, like the end of Line 9. Line 8 and Line 9 specify input and output ports of the other two instances of the **and** primitive. Line 10 is for instantiation of the **or** primitive at the output of the majority gate. The output of this gate is *y* that comes first in the port list, and is followed by inputs of the gate. In this example, intermediate signals for interconnection of gates are *im1*, *im2*, and *im3*. Scalar interconnecting wires need not be explicitly declared in Verilog.

```
`timescale 1ns/100ps          // Line 1
module maj3 ( a, b, c, y );
  input a, b, c;
  output y;

  and #(2,4)                  // Line 6
    ( im1, a, b ),            // Line 7
    ( im2, b, c ),            // Line 8
    ( im3, c, a );            // Line 9
  or  #(3,5) ( y, im1, im2, im3 );  // Line 10

endmodule
```

Figure 3.13 Verilog Code for the Majority Circuit

The three **and** instances could be written as three separate statements, like instantiation of the **or** primitive. If we were to specify different delay values for the three instances of the **and** primitive, we had to have three separate primitive instantiation statements.

Three-state gates are instantiated in the same way as the regular logic gates. Outputs of three-state gates can be wired to form wired-and, wired-or, or wiring logic. For various wiring functions, Verilog uses **wire**, **wand**, **wor**, **tri**, **tri0** and **tri1** net types. When two wires (**nets**) are connected, the resulting value depends on the two **net** values, as well as the type of the interconnecting **net**. Figure 3.14 shows **net** values for **net** types **wire**, **wand** and **wor**.

Driving **net** values									
	net values								
net type	00	01	0Z	0X	11	1Z	1X	ZZ	ZX	XX
wire	0	X	0	X	1	1	X	Z	X	X
wand	0	0	0	0	1	X	X	Z	X	X
wor	0	1	X	X	1	1	1	X	X	X

Figure 3.14 "net" Type Resolutions

The table shown in Figure 3.14 is called a **net** resolution table. Several examples of **net** resolutions are shown in Figure 3.15. The **tri net** type mentioned above is the same as the **wire** type. **tri0** and **tri1** types resolve to **0** and **1**, respectively, when driven by all **Z** values.

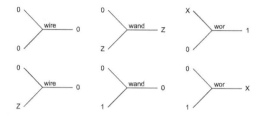

Figure 3.15 "net" Resolution Examples

Multiplexer Example. Figure 3.16 shows a 2-to-1 multiplexer using three-state gates. The Verilog code of this multiplexer is shown in Figure 3.17.

Lines 6 and 7 in Figure 3.17 instantiate two three-state gates. Their output is y, and since it is driven by both gates a wired-net is formed. Since y is not declared, its **net** type defaults to **wire**. When s is **1**, *bufif1* conducts and the value of b propagates to its output. At the same time, because s is **1**, *bufif0* does not conduct and its output becomes **Z**. Resolution of these values driving **net** y is determined by the **wire net** resolution as shown in Figure 3.14.

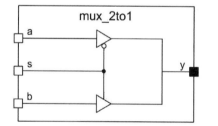

Figure 3.16 Multiplexer Using Three-state Gates

```
`timescale 1ns/100ps

module mux_2to1 ( a, b, s, y );
  input a, b, s;
  output y;
  bufif1 #(3) ( y, b, s );        // Line 6
  bufif0 #(5) ( y, a, s );        // Line 7
endmodule
```

Figure 3.17 Multiplexer Verilog Code

CMOS NAND Example.   As another example of instantiation of primitives, consider the two-input CMOS NAND gate shown in Figure 3.18.

The Verilog code of Figure 3.19 describes this CMOS NAND gate. Logically, NMOS transistors in a CMOS structure push **0** into the output of the gate. Therefore, in the Verilog code of the CMOS NAND, input to output direction of NMOS transistors are from *Gnd* towards *w*. Likewise, PMOS transistors push a **1** value into *w*, and therefore, their inputs are considered the *Vdd* node and their outputs are connected to the *w* node. The *im1* signal is an intermediate **net** and is explicitly declared.

In the Verilog code of CMOS NAND gate, primitive gate instance names are used. This naming (*T1, T2, T3, T4*) is optional for primitives and mandatory when modules are instantiated. Examples of module instantiations are shown in the next section.

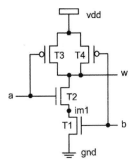

Figure 3.18 CMOS NAND Gate

```
module cmos_nand (a, b, w);
input a, b;
output w;
wire im1;
supply1 vdd;
supply0 gnd;

  nmos #(3, 4)
    T1 (im1, gnd, b),
    T2 (w, im1, a);

  pmos #(4, 5)
    T3 (w, vdd, a),
    T4 (w, vdd, b);

endmodule
```

Figure 3.19 CMOS NAND Verilog Description

3.2.2   Descriptions by Use of Equations

At a higher level than gates and transistors, a combinational circuit may be described by use of Boolean, logical, and arithmetic expressions. For this purpose the Verilog concurrent **assign** statement is used. Figure 3.20 shows Verilog operators that can be used with **assign** statements.

XOR Example. As our first example for using an **assign** statement consider the description of an XOR gate as shown in Figure 3.21. The **assign** statement uses y on the left-hand-side and equates it to Exclusive-OR of a, b, and c inputs.

Effectively, this **assign** statement is like driving y with the output of a 3-input **xor** primitive gate. The difference is that, the use of an **assign** statement gives us more flexibility and allows the use of more complex functions than what is available as primitive gates. Instead of being limited to the gates shown in Figure 3.11, we can write our own expressions using operators of Figure 3.20.

Bitwise Operators	&	|	^	~	~^	^~	
Reduction Operators	&	~&	|	~|	^	~^	^~
Arithmetic Operators	+	-	*	/	%		
Logical Operators	&&	||	!				
Compare Operators	<	>	<=	>=	==		
Shift Operators	>>	<<					
Concatenation Operators	{}	{ n{ } }					

Figure 3.20 Verilog Operators

```
module xor3 ( a, b, c, y );
   input a, b, c;
   output y;

   assign y = a ^ b ^ c;

endmodule
```

Figure 3.21 XOR Verilog Code

Full-Adder Example.   Figure 3.22 shows another example of using **assign** statements. This code corresponds to a full-adder circuit (see Chapter 2). The s output is the XOR result of a, b and ci inputs, and the co output is an AND-OR expression involving these inputs.

A delay value of 10 ns is used for the s output and 8 ns for the co output. As with the gate outputs, rise and fall delay values can be specified for a **net** that is used on the left-hand side of an **assign** statement. This construct allows the use of two delay values. If only one value is specified, it applies to both rise and fall transitions.

```
`timescale 1ns/100ps

module add_1bit ( a, b, ci, s, co );
  input a, b, ci;
  output s, co;
  assign #(10) s = a ^ b ^ ci;
  assign #(8) co = ( a & b ) | ( b & ci ) | ( a & ci );
endmodule
```

Figure 3.22 Full Adder Verilog Code

Another property of **assign** statements that also corresponds to gate instantiations is their concurrency. The statements in the Verilog module of Figure 3.22 are concurrent. This means that the order in which they appear in this module is not important. These statements are sensitive to events on their right hand sides. When a change of value occurs on any of the right hand side **net** or variables, the statement is evaluated and the resulting value is scheduled for the left hand side **net**.

Comparator Example. Figure 3.23 shows another example of using **assign** statements. This code describes a 4-bit comparator. The first **assign** statement uses a bitwise XOR operation on its right hand side. The result that is assigned to the *im* intermediate **net** is a 4-bit vector formed by XORing bits of a and b input vectors. The second **assign** statement uses the NOR reduction operator to NOR bits of *im* to generate the equal output for the 4-bit comparator.

The above describes the comparator using its Boolean function. However, using compare operators of Verilog, the *eq* output of the comparator may be written as:

```
assign eq = (a == b);
```

In this expression, $(a == b)$ results in **1** if a and b are equal, and **0** if they are not. This result is simply assigned to *eq*.

The right-hand side expression of an **assign** statement can have a condition expression using the **?** and **:** operators. These operators are like if-then-else. In reading expressions that involve a condition operator, **?** and **:** take places of **then** and **else** respectively. The if-condition appears to the left of **?**.

```
module comp_4bit ( a, b, eq );
  input [3:0]a, b;
  output eq;
  wire [3:0] im;
  assign im = a ^ b;
  assign eq = ~| im;
endmodule
```

Figure 3.23 Four-Bit Comparator

Multiplexer Example. Figure 3.24 shows a 2-to-1 multiplexer using a condition operator. The expression shown reads as follows: **if** s is **1, then** y is i1 **else** it becomes i0.

```
module mux2_1 ( i0, i1, s, y );
  input [3:0] i0, i1;
  input s;
  output [3:0] y;

  assign y = s ? i1 : i0;

endmodule
```

Figure 3.24 A 2-to-1 Mux Using Condition Operator

Decoder Example. Figure 3.25 shows another example using the condition operator. In this example a nesting of several **?:** operations are used to describe a decoder.

```
`timescale 1ns/100ps

module dcd2_4( a, b, d0, d1, d2, d3 );
  input a, b;
  output d0, d1, d2, d3;

  assign {d3, d2, d1, d0} =
    ( {a, b} == 2'b00 ) ? 4'b0001 :
    ( {a, b} == 2'b01 ) ? 4'b0010 :
    ( {a, b} == 2'b10 ) ? 4'b0100 :
    ( {a, b} == 2'b11 ) ? 4'b1000 :
      4'b0000;

endmodule
```

Figure 3.25 Decoder Using ?: and Concatenation

The decoder description also uses the concatenation operator { } to form vectors from its scalar inputs and outputs. The decoder has four outputs, d3, d2, d1 and d0 and two inputs a and b. Input values **00, 01, 10,** and **11** produce **0001, 0010, 0100,** and **1000** outputs. In order to be able to compare a and b with their possible values, a two-bit vector is formed by concatenating a and b. The {a, b} vector is then compared with the four possible values it can take using a nesting of **?:** operations.

Similarly, in order to be able to place vector values on the outputs, the four outputs are concatenated using the { } operator and used on the left-hand side of the **assign** statement shown in Figure 3.25.

This example also shows the use of sized numbers. Constants for the inputs and outputs have the general format of **n`bm**. In this format, **n** is the number of bits, **b** is the base specification and **m** is the number in base **b**. For calculation of the corresponding constant, number **m** in base **b** is translated to **n** bit binary. For example, 4`hA becomes 1010 in binary.

Adder Example. For another example using **assign** statements, consider an 8-bit adder circuit with a carry-in and a carry-out output. The Verilog code of this adder, shown in Figure 3.26, uses an **assign** statement to set concatenation of co on the left-hand side of s to the sum of a, b and ci. This sum results in nine bits with the left-most bit being the resulting carry. The sum is captured in the 9-bit left-hand side of the **assign** statement in {co, s}.

So far in this section we have shown the use of operators of Figure 3.20 in **assign** statements. A Verilog description may contain any number of **assign** statements and can use any mix of the operators discussed. The next example shows multiple **assign** statements.

```
module add_4bit ( a, b, ci, s, co );
  input [7:0] a, b;
  output [7:0] s;
  input ci;
  output co;

  assign {co, s} = a + b + ci;

endmodule
```

Figure 3.26 Adder with Carry-in and Carry-out

ALU Example. As our final example of **assign** statements, consider an ALU that performs add and subtract operations and has two flag outputs gt and $zero$. The gt output becomes **1** when input a is greater than input b, and the $zero$ output becomes **1** when the result of the operation performed by the ALU is **0**.

Figure 3.27 shows the Verilog code of this ALU. Used in this description are arithmetic, concatenation, condition, compare and relational operations.

```
module ALU ( a, b, ci, addsub, gt, zero, co, r );
  input [7:0] a, b;
  output [7:0] r;
  input ci;
  output gt, zero, co;

  assign {co, s} = addsub ? (a + b + ci) : (a - b - ci);
  assign gt = (a>b);
  assign zero = (r == 0);

endmodule
```

Figure 3.27 ALU Verilog Code Using a Mix of Operations

3.2.3 Descriptions with Procedural Statements

At a higher level of abstraction than describing hardware with gates and expressions, Verilog provides constructs for procedural description of hardware. Unlike gate instantiations and **assign** statements that correspond to concurrent sub-structures of a hardware component, procedural statements describe the hardware by its behavior. Also, unlike concurrent statements that appear directly in a module body, procedural statements must be enclosed in procedural blocks before they can be put inside a module.

The main procedural block in Verilog is the **always** block. This is considered a concurrent statement that runs concurrent with all other statements in a module. Within this statement, procedural statements like **if-else** and **case** statements are used and are executed sequentially. If there are more than one procedural statement inside a procedural block, they must be bracketed by **begin** and **end** keywords.

Unlike assignments in concurrent bodies that model driving logic for left hand side wires, assignments in procedural blocks are assignments of values to variables that hold their assigned values until a different value is assigned to them. A variable used on the left hand side of a procedural assignment must be declared as **reg**.

An event control statement is considered a procedural statement, and is used inside an **always** block. This statement begins with an at-sign, and in its simplest form, includes a list of variables in the set of parenthesis that follow the at-sign, e.g., @ (v1 **or** v2 ...); .

When the flow of the program execution within an **always** block reaches an event-control statement, the execution halts (suspends) until an event occurs on one of the variables in the enclosed list of variables. If an event-control statement appears at the beginning of an **always** block, the variable list it contains is referred to as the *sensitivity list* of the **always** block. For combinational circuit modeling all variables that are read inside a procedural block must appear on its sensitivity list.

Examples that follow show various ways combinational component may be modeled by procedural blocks.

Majority Example. Figure 3.28 shows a majority circuit described by use of an **always** block. In the declarative part of the module shown, the y output is declared as **reg** since this variable is to be assigned a value inside a procedural block.

The **always** block describing the behavior of this circuit uses an event control statement that encloses a list of variables that is considered as the sensitivity list of the **always** block. The **always** block is said to be sensitive to a, b and c variables. When an event occurs on any of these variables, the flow into the **always** block begins and as a result, the result of the Boolean expression shown will be assigned to variable y. This variable holds its value until the next time an event occurs on a, b, or c inputs.

In this example, since the **begin** and **end** bracketing only includes one statement, its use is not necessary. Furthermore, the syntax of Verilog allows elimination of semicolon after an event control statement. This effectively collapses the event control and the statement that follows it into one statement.

```
module maj3 ( a, b, c, y );
  input a, b, c;
  output y;
  reg y;

  always @( a or b or c )
  begin
    y = (a & b) | (b &c) | (a & c);
  end

endmodule
```

Figure 3.28 Procedural Block Describing a Majority Circuit

Majority Example with Delay. The Verilog code shown in Figure 3.29 is a majority circuit with a 5ns delay. Following the **always** keyword, the statements in this procedural block are an event-control, a delay-control and a procedural assignment. The delay-control statement begins with a sharp-sign and is followed by a delay value. This statement causes the flow into this procedural block to be suspended for 5ns. This means that after an event on one of the circuit inputs, evaluation and assignment of the output value to y takes place after 5 nanoseconds.

Note in the description of Figure 3.29 that **begin** and **end** bracketing is not used. As with the event-control statement, a delay-control statement can collapse into its next statement by removing their separating semicolon. The event-control, delay-control and assignment to y become a single procedural statement in the **always** block of *maj3* code.

```
`timescale 1ns/100ps

module maj3 ( a, b, c, y );
  input a, b, c;
  output y;
  reg y;

  always @( a or b or c ) #5 y = (a & b) | (b &c) | (a & c);

endmodule
```

Figure 3.29 Majority Gate with Delay

Full-Adder Example. Another example of using procedural assignments in a procedural block is shown in Figure 3.30. This example describes a full-adder with sum and carry-out outputs.

The **always** block shown is sensitive to a, b, and ci inputs. This means that when an event occurs on any of these inputs, the **always** block wakes up and executes all its statements in the order that they appear. Since assignments to s and co outputs are procedural, both these outputs are declared as **reg**.

The delay mechanism used in the full-adder of Figure 3.30 is called an intra-statement delay that is different than that of the majority circuit of Figure 3.29.

```
`timescale 1ns/100ps

module add_1bit ( a, b, ci, s, co );
  input a, b, ci;
  output s, co;
  reg s, co;

  always @( a or b or ci )
  begin
    s = #5 a ^ b ^ ci;
    co = #3 (a & b) | (b &ci) | (a & ci);
  end
endmodule
```

Figure 3.30 Full-Adder Using Procedural Assignments

In the majority circuit, the delay simply delays execution of its next statement. However, the intra-statement delay of Figure 3.30 only delays the assignment of the calculated value of the right-hand side to the left-hand side variable. This means that in Figure 3.30, as soon as an event occurs on an input, the expression $a \wedge b \wedge c$ is evaluated. But, the assignment of the evaluated value to s and proceeding to the next statement takes 5ns.

Because assignment to co follows that to s, the timing of the former depends on that of the latter, and evaluation of the right-hand side of co begins 5ns after an input change. Therefore, co receives its value 8ns after an input change occurs. To remove this timing dependency and be able to define the timing of each statement independent of its previous one, a different kind of assignment must be used.

```
`timescale 1ns/100ps

module add_1bit ( a, b, ci, s, co );
  input a, b, ci;
  output s, co;
  reg s, co;

  always @( a or b or ci )
  begin
    s <= #5 a ^ b ^ ci;
    co <= #8 (a & b) | (b &ci) | (a & ci);
  end
endmodule
```

Figure 3.31 Full-Adder Using Non-Blocking Assignments

Assignments in Figure 3.30 are of the blocking type. Such statements block the flow of the program until they are completed. A different assignment is of the non-blocking type. A different version of the full-adder that uses this construct is shown in Figure 3.31. This assignment schedules its right hand side value into its left hand side to take place after the specified delay. Program flow continues into the next statement while propagation of values into the first left hand side is still going on.

In the example of Figure 3.31, evaluation of the right hand side of s is done immediately after an input changes. Evaluation of the right hand side of co occurs 8ns after that. To make s and co delays match those of Figure 3.30, an 8 nanoseconds delay is used for assignment to co.

Since our focus is on synthesizable coding and gate delay timing issues are not of importance, we will mostly use blocking assignments in this book.

Procedural Multiplexer Example. For another example of a procedural block, consider the 2-to-1 multiplexer of Figure 3.32. This example uses an **if-else** construct to set y to i0 or i1 depending on the value of s.

As in the previous examples, all circuit variables that participate in determination of value of y appear on the sensitivity list of the **always** block. Also since y appears on the left hand side of a procedural assignment, it is declared as **reg**.

The **if-else** statement shown in Figure 3.32 has a condition part that uses an equality operator. If the condition is false (or equal to **0**), the block of statements that follow it will be taken, otherwise block of statements after the **else** are taken. In both cases, the block of statements must be bracketed by **begin** and **end** keywords if there is more than one statement in a block.

```
module mux2_1 ( i0, i1, s, y );
  input i0, i1, s;
  output y;
  reg y;

  always @( i0 or i1 or s )  begin
    if ( s==1'b0 )
      y = i0;
    else
      y = i1;
  end
endmodule
```

Figure 3.32 Procedural Multiplexer

Procedural ALU Example. The **if-else** statement, used in the previous example, is easy to use, descriptive and expandable. However, when many choices exist, a **case**-statement which is more structured may be a better choice. The ALU description of Figure 3.33 uses a **case** statement to describe an ALU with add, subtract, AND and XOR functions.

```
module alu_4bit ( a, b, f, y );
  input [3:0] a, b;
  input [1:0] f;
  output [3:0] y;
  reg [3:0] y;

  always @ ( a or b or f )  begin
    case ( f )
      2'b00 : y = a + b;
      2'b01 : y = a - b;
      2'b10 : y = a & b;
      2'b11 : y = a ^ b;
      default: y = 4'b0000;
    endcase
  end
endmodule
```

Figure 3.33 Procedural ALU

The ALU has a and b data inputs and a 2-bit f input that selects its function.   The Verilog code shown in Figure 3.33 uses a, b and f on its sensitivity list. The **case**-statement shown in the **always** block uses f to select one of the **case** alternatives. The last alternative is the **default** alternative that is taken when f does not match any of the alternatives that appear before it. This is necessary to make sure that unspecified input values (here, those that contain **X** and/or **Z**) cause the assignment of the default value to the output and not leave it unspecified.

3.2.4   Combinational Rules

Completion of **case** alternatives or **if-else** conditions is an important issue in combinational circuit coding. In an **always** block, if there are conditions under which the output of a combinational circuit is not assigned a value, because of the property of **reg** variables the output retains its old value. The retaining of old value infers a latch on the output. Although, in some designs this latching is intentional, obviously it is unwanted when describing combinational circuits. With this, we have set two rules for coding combinational circuits with **always** blocks.

1. List all inputs of the combinational circuit in the sensitivity list of the **always** block describing it.
2. Make sure all combinational circuit outputs receive some value regardless of how the program flows in the conditions of **if-else** and/or **case** statements.   If there are too many conditions to check, set all outputs to their inactive values at the beginning of the **always** block.

3.2.5   Bussing

Bus structures can be implemented by use of multiplexers or three-state logic. In Verilog, various methods of describing combinational circuits can be used for the description of a bus.

Figure 3.34 shows Verilog coding of *busout* that is a three-state bus and has three sources, *busin1*, *busin2*, and *busin3*. Sources of *busout* are put on this bus by active-high enabling control signals, *en1*, *en2* and *en3*. Using the value of an enabling signal, a condition statement either selects a bus driver or a 4-bit **Z** value to drive the *busout* output.

```
module bussing (busin1, busin2, busin3, en1, en2, en3, busout );
   input [3:0] busin1, busin2, busin3;
   input en1, en2, en3;
   output [3:0] busout;

   assign busout = en1 ? busin1 : 4'bzzzz;
   assign busout = en2 ? busin2 : 4'bzzzz;
   assign busout = en3 ? busin3 : 4'bzzzz;

endmodule
```

Figure 3.34 Implementing a 3-State Bus

Verilog allows multiple concurrent drivers for **net**s. However, a variable declared as a **reg** and used on a left hand side in a procedural block (**always** block), can only be driven by one source. This makes the use of **net**s more appropriate for representing busses.

3.3   Sequential Circuits

As with any digital circuit, a sequential circuit can be described in Verilog by use of gates, Boolean expressions, or behavioral constructs (e.g., the **always** statement). While gate level descriptions enable a more detailed description of timing and delays, because of complexity of clocking and register and flip-flop controls, these circuits are usually described by use of procedural **always** blocks. This section shows various ways sequential circuits are described in Verilog. The following discusses primitive structures like latch and flip-flops, and then generalizes coding styles used for representing these structures to more complex sequential circuits including counters and state machines.

3.3.1   Basic Memory Elements at the Gate Level

A clocked D-latch latches its input data during an active clock cycle. The latch structure retains the latched value until the next active clock cycle. This element is the basis of all static memory elements.

A simple implementation of the D-latch that uses cross-coupled NOR gates is shown in Figure 3.35. The Verilog code of Figure 3.36 corresponds to this D-latch circuit. This description uses primitive **and** and **nor** structures.

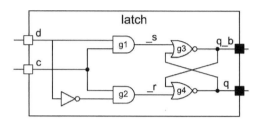

Figure 3.35 Clocked D-latch

```
`timescale 1ns/100ps

module latch ( d, c, q, q_b );
  input d, c;
  output q, q_b;
  wire _s, _r;

    and  #(6)  g1 ( _s, c, d ),
               g2 ( _r, c, ~d );
    nor  #(4)  g3 ( q_b, _s, q ),
               g4 ( q, _r, q_b );
endmodule
```

Figure 3.36 Verilog Code for a Clocked D-latch

As shown in this Verilog code, the tilde (~) operator is used to generate the complement of the d input of the latch. Using AND gates, the d input and its complement are gated to generate internal $\_s$ and $\_r$ inputs. These are inputs to the cross-coupled NOR structure that is the core of the memory in this latch.

Alternatively, the same latch can be described with an **assign** statement as shown below.

    assign #(3) q = c ? d : q;

This statement simply describes what happens in a latch. The statement says that when c is **1**, the q output receives d, and when c is **0** it retains its old value. Using two such statements with complementary clock values describe a master-slave flip-flop. As shown in Figure 3.37, the qm **net** is the master output and q is the flip-flop output.

```
`timescale 1ns/100ps

module master_slave ( d, c, q );
  input d, c;
  output q;

  wire qm;

  assign #(3) qm = c ? d : qm;
  assign #(3) q = ~c ? qm : q;

endmodule
```

Figure 3.37 Master-Slave Flip-Flop

This code uses two concurrent **assign** statements. As discussed before, these statements model logic structures with **net** driven outputs (qm and q). The order in which the statements appear in the body of the *master_slave* **module** is not important.

3.3.2   Memory Elements Using Procedural Statements

Although latches and flip-flops can be described by primitive gates and **assign** statements, such descriptions are hard to generalize, and describing more complex register structures cannot be done this way. This section uses **always** statements to describe latches and flip-flops. We will show that the same coding styles used for these simple memory elements can be generalized to describe memories with complex control as well as functional register structures like counters and shift-registers.

```
module latch ( d, c, q, q_b );
  input d, c;
  output q, q_b;
  reg q, q_b;

  always @ ( c or d )
    if ( c ) begin
      #4 q = d;
      #3 q_b = ~d;
    end
endmodule
```

Figure 3.38 Procedural Latch

Latches. Figure 3.38 shows a D-latch described by an **always** statement. The outputs of the latch are declared as **reg** because they are being driven inside the **always** procedural block. Latch clock and data inputs (c and d) appear in the sensitivity list of the **always** block, making this procedural statement sensitive to c and d. This means that when an event occurs on c or d, the

always block wakes up and it executes all its statements in the sequential order from begin to end.

The **if**-statement enclosed in the **always** block puts d into q when c is active. This means that if c is **1** and d changes, the change on d propagates to the q output. This behavior is referred to as transparency, which is how latches work. While clock is active, a latch structure is transparent, and input changes affect its output.

Any time the **always** statement wakes up, if c is **1**, it waits 4 nanoseconds and then puts d into q. It then waits another 3 nanoseconds and then puts the complement of d into q_b. This makes the delay of the q_b output 7 ns.

D Flip-Flop. While a latch is transparent, a change on the D-input of a D flip-flops does not directly pass on to its output. The Verilog code of Figure 3.39 describes a positive-edge trigger D-type flip-flop.

The sensitivity list of the procedural statement shown includes **posedge** of *clk*. This **always** statement only wakes up when *clk* makes a **0** to **1** transition. When this statement does wake up, the value of d is put into q. Obviously this behavior implements a rising-edge D flip-flop.

```
`timescale 1ns/100ps

module d_ff ( d, clk, q, q_b );
   input d, clk;
   output q, q_b;
   reg q, q_b;

   always @ ( posedge clk )
   begin
      #4 q = d;
      #3 q_b = ~d;
   end
endmodule
```

Figure 3.39 A Positive-Edge D Flip-Flop

Instead of **posedge**, use of **negedge** would implement a falling-edge D flip-flop. After the specified edge, the flow into the **always** block begins. In our description, this flow is halted by 4 nanoseconds by the #4 delay-control statement. After this delay, the value of d is read and put into q. Following this transaction, the flow into the **always** block is again halted by 3 nanoseconds, after which $\sim d$ is put into qb. This makes the delay of q after the edge of the clock equal to 4 nanoseconds. The delay for q_b becomes the accumulation of the delay values shown, and it is 7 nanoseconds. Delay values are ignored in synthesis.

Synchronous Control. The coding style presented for the above simple D flip-flop is a general one and can be expanded to cover many features found in flip-flops and even memory structures. The description shown in Figure 3.40 is a D-type flip-flop with synchronous set and reset (s and r) inputs.

The description uses an **always** block that is sensitive to the positive-edge of *clk*. When *clk* makes a **0** to **1** transition, the flow into the **always** block begins. Immediately after the positive-edge, *s* is inspected and if it is active (**1**), after 4 ns *q* is set to **1** and 3 ns after that *q_b* is set to **0**. Following the positive-edge of *clk*, if *s* is not **1**, *r* is inspected and if it is active, *q* is set to **0**. If neither *s* nor *r* are **1**, the flow of the program reaches the last **else** part of the **if**-statement and assigns *d* to *q*.

The behavior discussed here only looks at *s* and *r* on the positive-edge of *clk*, which corresponds to a rising-edge trigger D-type flip-flop with synchronous active high set and reset inputs. Furthermore, the set input is given a higher priority over the reset input. The flip-flop structure that corresponds to this description is shown in Figure 3.41.

Other synchronous control inputs can be added to this flip-flop in a similar fashion. A clock enable (*en*) input would only require inclusion of an **if**-statement in the last **else** part of the **if**-statement in the code of Figure 3.40.

```
module d_ff ( d, s, r, clk, q, q_b );
  input d, clk, s, r;
  output q, q_b;
  reg q, q_b;

  always @ ( posedge clk ) begin
    if ( s ) begin
      #4 q = 1'b1;
      #3 q_b = 1'b0;
    end else if ( r ) begin
      #4 q = 1'b0;
      #3 q_b = 1'b1;
    end else begin
      #4 q = d;
      #3 q_b = ~d;
    end
  end

endmodule
```

Figure 3.40 D Flip-Flop with Synchronous Control

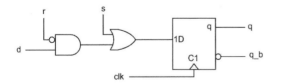

Figure 3.41 D Flip-Flop with Synchronous Control

Asynchronous Control. The control inputs of the flip-flop of Figure 3.40 are synchronous because the flow into the **always** statement is only allowed to start when the **posedge** of *clk* is observed. To change this to a flip-flop with asynchronous control, it is only required to include asynchronous control inputs in the sensitivity list of its procedural statement.

Figure 3.42 shows a D flip-flop with active high asynchronous set and reset control inputs. Note that the only difference between this description and the code of Figure 3.40 (synchronous control) is the inclusion of **posedge** *s* and **posedge** *r* in the sensitivity list of the **always** block. This inclusion allows the flow into the procedural block to begin when *clk* becomes **1** or *s* becomes **1** or *r* becomes **1**. The **if**-statement in this block checks for *s* and *r* being **1**, and if none are active (activity levels are high) then clocking *d* into *q* occurs.

An active high (low) asynchronous input requires inclusion of **posedge** (**negedge**) of the input in the sensitivity list, and checking its **1** (**0**) value in the **if**-statement in the **always** statement. Furthermore, clocking activity in the flip-flop (assignment of *d* into *q*) must always be the last choice in the **if**-statement of the procedural block.

The graphic symbol corresponding to the flip-flop of Figure 3.42 is shown in Figure 3.43.

```
module d_ff ( d, s, r, clk, q, q_b );
  input d, clk, s, r;
  output q, q_b;
  reg q, q_b;
  always @ (posedge clk or posedge s or posedge r )
  begin
    if ( s ) begin
      #4 q = 1'b1;
      #3 q_b = 1'b0;
    end else if ( r ) begin
      #4 q = 1'b0;
      #3 q_b = 1'b1;
    end else begin
      #4 q = d;
      #3 q_b = ~d;
    end
  end
endmodule
```

Figure 3.42 D Flip-Flop with Asynchronous Control

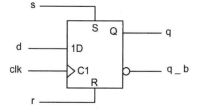

Figure 3.43 Flip-Flop with Asynchronous Control Inputs

3.3.3   Registers, Shifters and Counters

Registers, shifter-registers, counters and even sequential circuits with more complex functionalities can be described by simple extensions of the coding styles presented for the flip-flops. In most cases, the functionality of the circuit only affects the last **else** of the **if**-statement in the procedural statement of codes shown for the flip-flops.

Registers. Figure 3.44 shows an 8-bit register with synchronous set and reset inputs. The *set* input puts all **1**s in the register and the *reset* input resets it to all **0**s. The main difference between this and the flip-flop with synchronous control is the vector declaration of inputs and outputs.

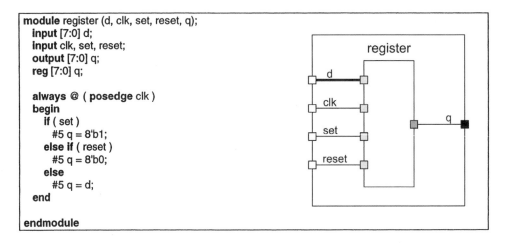

```
module register (d, clk, set, reset, q);
  input [7:0] d;
  input clk, set, reset;
  output [7:0] q;
  reg [7:0] q;

  always @ ( posedge clk )
  begin
    if ( set )
      #5 q = 8'b1;
    else if ( reset )
      #5 q = 8'b0;
    else
      #5 q = d;
  end

endmodule
```

Figure 3.44 An 8-bit Register

Shift-Registers. A 4-bit shift-register with right- and left-shift capabilities, a serial-input, synchronous reset input, and parallel loading capability is shown in Figure 3.45. As shown, only the positive-edge of *clk* is included in the sensitivity list of the **always** block of this code, which makes all activities of the shift-register synchronous with the clock input. If *rst* is **1**, the register is reset, if *ld* is **1** parallel *d* inputs are loaded into the register, and if none are **1** shifting left or right takes place depending on the value of the *l_r* input (**1** for left, **0** for right). Shifting in this code is done by use of the concatenation operator { }. For left-shift, *s_in* is concatenated to the right of *q[2:0]* to form a 4-bit vector that is put into *q*. For right-shift, *s_in* is concatenated to the left of *q[3:1]* to form a 4-bit vector that is clocked into *q[3:0]*.

The style used for coding this register is the same as that used for flip-flops and registers presented earlier. In all these examples, a single procedural block handles function selection (e.g., zeroing, shifting, or parallel loading) as well as clocking data into the register output.

```
module shift_reg (d, clk, ld, rst, l_r, s_in, q);
    input [3:0] d;
    input clk, ld, rst, l_r, s_in;
    output [3:0] q;
    reg [3:0]q;

    always @( posedge clk ) begin
        if ( rst )
            #5 q = 4'b0000;
        else if ( ld )
            #5 q = d;
        else if ( l_r )
            #5 q = {q[2:0], s_in};
        else
            #5 q = {s_in, q[3:1]};
    end

endmodule
```

Figure 3.45 A 4-bit Shift Register

Another style of coding registers, shift-registers and counters is to use a combinational procedural block for function selection and another for clocking.

As an example, consider a shift-register that shifts s_cnt number of places to the right or left depending on its sr or sl control inputs (Figure 3.46). The shift-register also has an ld input that enables its clocked parallel loading. If no shifting is specified, i.e., sr and sl are both zero, then the shift register retains its old value.

The Verilog code of Figure 3.46 shows two procedural blocks that are identified by *combinational* and *register*. A block name appears after the **begin** keyword that begins a block and is separated from this keyword by use of a colon. Figure 3.47 shows a graphical representation of the coding style used for the description of our shifter.

The *combinational* block is sensitive to all inputs that can affect the shift register output. These include the parallel d_in, the s_cnt shift-count, sr and sl shift control inputs, and the ld load control input. In the body of this block an **if-else** statement decides on the value placed on the int_q internal variable. The value selection is based on values of ld, sr, and sl. If ld is **1**, int_q becomes d_in that is the parallel input of the shift register. If sr or sl is active, int_q receives the previous value of int_q shifted to right or left as many as s_cnt places. In this example, shifting is done by use of the >> and << operators. On the left, these operators take the vector to be shifted, and on the right they take the number of places to shift.

The int_q variable that is being assigned values in the *combinational* block is a 4-bit **reg** that connects the output of this block to the input of the register block.

The *register* block is a sequential block that handles clocking int_q into the shift register output. This block (as shown in Figure 3.46) is sensitive to the positive edge of clk and its body consists of a single **reg** assignment.

Note in this code that both q and int_q are declared as **reg** because they are both receiving values in procedural blocks.

```
module shift_reg ( d_in, clk, s_cnt, sr, sl, ld, q );
   input [3:0] d_in;
   input clk, sr, sl, ld;
   input [1:0] s_cnt;
   output [3:0] q;
   reg [3:0] q, int_q;

   always @ ( d_in or s_cnt or sr or sl or ld ) begin: combinational
      if ( ld ) int_q = d_in;
      else if ( sr ) int_q = int_q >> s_cnt;
      else if ( sl ) int_q = int_q << s_cnt;
      else int_q = int_q;
   end

   always @ ( posedge clk ) begin: register
      q = int_q;
   end

endmodule
```

Figure 3.46 Shift-Register Using Two Procedural Blocks

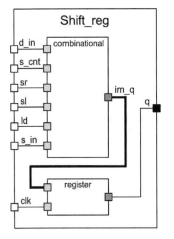

Figure 3.47 Shifter Block Diagram

Counters. Any of the styles described for the shift-registers in the previous discussion can be used for describing counters. A counter counts up or down, while a shift-register shifts right or left. We use arithmetic operations in counting as opposed to shift or concatenation operators in shift-registers.

Figure 3.48 shows a 4-bit up-down counter with a synchronous *rst* reset input. The counter has an *ld* input for doing the parallel loading of *d_in* into the counter. The counter output is *q* and it is declared as **reg** since it is receiving values within a procedural statement.

Discussion about synchronous and asynchronous control of flip-flops and registers also apply to the counter circuits. For example, inclusion of ***posedge*** *rst* in the sensitivity list of the counter of Figure 3.48 would make its resetting asynchronous.

```
module counter (d_in, clk, rst, ld, u_d, q );
    input [3:0] d_in;
    input clk, rst, ld, u_d;
    output [3:0] q;
    reg [3:0] q;

    always @ ( posedge clk ) begin
        if ( rst )
            q = 4'b0000;
        else if ( ld )
            q = d_in;
        else if ( u_d )
            q = q + 1;
        else
            q = q - 1;
    end

endmodule
```

Figure 3.48 An Up-Down Counter

3.3.4   State Machine Coding

Coding styles presented so far can be further generalized to cover finite state machines of any type. This section shows coding for Moore and Mealy state machines. The examples we will use are simple sequence detectors. These circuits represent the controller part of a digital system that has been partitioned into a data path and a controller. The coding styles used here apply to such controllers, and will be used in later chapters of this book to describe CPU and multiplier controllers.

Moore Detector. State diagram for a Moore sequence detector detecting **101** on its *x* input is shown in Figure 3.49. The machine has four states that are labeled, *reset, got1, got10,* and *got101*. Starting in *reset,* if the **101** sequence is detected, the machine goes into the *got101* state in which the output becomes **1**. In addition to the *x* input, the machine has a *rst* input that forces the machine into its *reset* state. The resetting of the machine is synchronized with the clock.

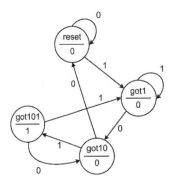

Figure 3.49 A Moore Sequence Detector

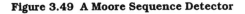

```
module moore_detector (x, rst, clk, z );
  input x, rst, clk;
  output z;  reg z;
  parameter [1:0]  reset = 0, got1 = 1, got10 = 2, got101 = 3;
  reg [1:0] current;
  always @ ( posedge clk ) begin
   if ( rst ) begin
    current = reset;  z = 1'b0;
   end
   else case ( current )
    reset:  begin
        if ( x==1'b1 ) current = got1;
        else current = reset;  z = 1'b0;
       end
    got1:  begin
        if ( x==1'b0 ) current = got10;
        else current = got1; z = 1'b0;
       end
    got10:  begin
        if ( x==1'b1 ) begin
          current = got101; z=1'b1;
        end else begin
          current = reset;  z = 1'b0;
        end
       end
    got101:  begin
        if ( x==1'b1 ) current = got1;
        else current = got10;
        z = 1'b0;
       end
   endcase
  end
endmodule
```

Figure 3.50 Moore Machine Verilog Code

The Verilog code of the Moore machine of Figure 3.49 is shown in Figure 3.50. After the declaration of inputs and outputs of this module, **parameter** declaration declares four states of the machine as two-bit parameters. The square-brackets following the **parameter** keyword specify the size of parameters being declared. Following parameter declarations in the code of Figure 3.50, the two-bit *current* **reg** type variable is declared. This variable holds the current state of the state machine.

The **always** block used in the module of Figure 3.50 describes state transitions and output assignments of the state diagram of Figure 3.49. The main task of this procedural block is to inspect input conditions (values on *rst* and *x*) during the present state of the machine defined by *current* and set values into *current* for the next state of the machine.

The flow into the **always** block begins with the positive edge of *clk*. Since all activities in this machine are synchronized with the clock, only *clk* appears on the sensitivity list of the **always** block. Upon entry into this block, the *rst* input is checked and if it is active, *current* is set to *reset* (*reset* is a declared parameter and its value is **0**). The value put into *current* in this pass through the **always** block gets checked in the next pass with the next edge of the clock. Therefore this assignment is regarded as the next-state assignment. When this assignment is made, the **if-else** statements skip the rest of the code of the **always** block, and this **always** block will next be entered with the next positive edge of *clk*.

Upon entry into the **always** block, if *rst* is not **1**, program flow reaches the **case** statement that checks the value of *current* against the four states of the machine. Figure 3.51 shows an outline of this **case**-statement.

```
case ( current )
    reset:  begin . . . end
    got1:   begin . . . end
    got10:  begin . . . end
    got101: begin  . . . end
endcase
```

Figure 3.51 case-Statement Outline

The **case**-statement shown has four **case**-alternatives. A **case**-alternative is followed by a block of statements bracketed by the **begin** and **end** keywords. In each such block, actions corresponding to the active state of the machine are taken.

Figure 3.52 shows the Verilog code of the *got10* state and its diagram from the state diagram of Figure 3.49. As shown here, the **case**-alternative that corresponds to the *got10* state only specifies the next values for the state and output of the circuit.

Note, for example, that the Verilog code segment of state *got10* does not specify the output of this state. Instead, the next value of *current* and the next value of *z* are specified based on the value of *x*. If *x* is **1**, the next state becomes *got101* in which *z* is **1**, and if *x* is **0**, the next state becomes *reset*.

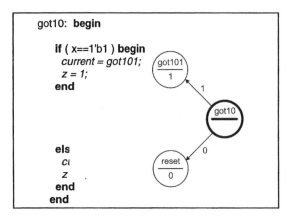

Figure 3.52 Next Values from *got10*

In this coding style, for every state of the machine there is a **case**-alternative that specifies the next state values. For larger machines, there will be more **case**-alternatives, and more conditions within an alternative. Otherwise, this style can be applied to state machines of any size and complexity.

This same machine can be described in Verilog in many other ways. We will show alternative styles of coding state machines by use of examples that follow.

A Mealy Machine Example. Unlike a Moore machine that has outputs that are only determined by the current state of the machine, in a Mealy machine, the outputs are determined by the state the machine is in as well as the inputs of the circuit. This makes Mealy outputs not fully synchronized with the circuit clock. In the state diagram of a Mealy machine the outputs are specified along the edges that branch out of the states of the machine.

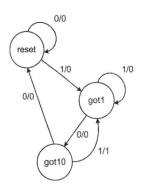

Figure 3.53 A 101 Mealy Detector

Figure 3.53 shows a **101** Mealy detector. The machine has three states, *reset, got1* and *got10*. While in *got10*, if the *x* input becomes **1** the machine prepares to go to its next state with the next clock. While waiting for the clock, its output becomes **1**. While on the edge that takes the machine out of *got10*, if the clock arrives the machine goes into the *got1* state. This machine allows overlapping sequences. The machine has no external resetting mechanism. A sequence of two zeros on input *x* puts the machine into the *reset* state in a maximum of two clocks.

The Verilog code of the **101** Mealy detector is shown in Figure 3.54. After input and output declarations, a **parameter** declaration defines bit patterns (state assignments) for the states of the machine. Note here that state value 3 or **11** is unused. As in the previous example, we use the *current* two-bit **reg** to hold the current state of the machine.

After the declarations, an **initial** block sets the initial state of the machine to *reset*. This procedure for initializing the machine is only good for simulation and is not synthesizable.

This example uses an **always** block for specifying state transitions and a separate statement for setting values to the *z* output. The **always** statement responsible for state transitions is sensitive to the circuit clock and has a **case** statement that has **case** alternatives for every state of the machine. Consider for example, the *got10* state and its corresponding Verilog code segment, as shown in Figure 3.55.

```
module mealy_detector ( x, clk, z );
  input x, clk;
  output z;
  parameter [1:0]
   reset  = 0, // 0 = 0 0
   got1   = 1, // 1 = 0 1
   got10  = 2; // 2 = 1 0

  reg [1:0] current;

  initial current = reset;
  always @ ( posedge clk )
  begin
   case ( current )
    reset:  if( x==1'b1 ) current = got1;
       else current = reset;
    got1:  if( x==1'b0 ) current = got10;
       else current = got1;
    got10:  if( x==1'b1 ) current = got1;
        else current = reset;
     default: current = reset;
    endcase
  end
  assign z= ( current==got10 && x==1'b1 ) ? 1'b1 : 1'b0;

endmodule
```

Figure 3.54 Verilog Code of 101 Mealy Detector

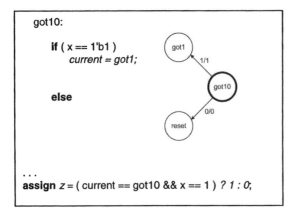

Figure 3.55 Coding a Mealy State

As shown, the Verilog code of this state only specifies its next states and does not specify the output values. Notice also in this code segment that the **case** alternative shown does not have **begin** and **end** bracketing. Actually, **begin** and **end** keywords do not appear in blocks following **if** and **else** keywords either.

Verilog only requires **begin** and **end** bracketing if there is more than one statement in a block. The use of this bracketing around one statement is optional. Since the **if** part and the **else** part each only contain one statement, **begin** and **end** keywords are not used. Furthermore, since the entire **if-else** statement reduces to only one statement, the **begin** and **end** keywords for the **case**-alternative are also eliminated.

The last **case**-alternative shown in Figure 3.54 is the **default** alternative. When checking *current* against all alternatives that appear before the **default** statement fail, this alternative is taken. There are several reasons that we use this default alternative.   One is that, our machine only uses three of the possible four 2-bit assignments and **11** is unused. If the machine ever begins in this state, the default case makes *reset* the next state of the machine. The second reason why we use **default** is that Verilog assumes a four-value logic system that includes **Z** and **X**. If *current* ever contains a **Z** or **X**, it does not match any of the defined case alternatives, and the default case is taken. Another reason for use of **default** is that our machine does not have a hard reset and we are making provisions for it to go to the *reset* state. The last reason for **default** is that it is just a good idea to have it.

The last statement in the code fragment of Figure 3.55 is an **assign** statement that sets the z output of the circuit. This statement is a concurrent statement and is independent of the **always** statement above it. When *current* or x changes, the right hand side of this assignment is evaluated and a value of **0** or **1** is assigned to z. Conditions on the right hand side of this assignment are according to values put in z in the state diagram of Figure 3.54. Specifically, the output is **1** when *current* is *got10* and x is **1**, otherwise it is **0**. This statement implements a combinational logic structure with *current* and x inputs and z output.

Huffman Coding Style. The Huffman model for a digital system characterizes it as a combinational block with feedbacks through an array of registers. Verilog coding of digital systems according to the Huffman model uses an **always** statement for describing the register part and another concurrent statement for describing the combinational part.

We will describe the state machine of Figure 3.49 to illustrate this style of coding. Figure 3.56 shows the combinational and register part partitioning that we will use for describing this machine. The *combinational* block uses x and p_state as input and generates z and n_state. The *register* block clocks n_state into p_state, and reset p_state when rst is active.

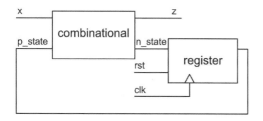

Figure 3.56 Huffman Partitioning of 101 Moore Detector

Figure 3.57 shows the Verilog code of Figure 3.49 according to the partitioning of Figure 3.56. As shown, parameter declaration declares the states of the machine. Following this declaration, n_state and p_state variables are declared as two-bit **reg**s that hold values corresponding to the states of the **101** Moore detector. The *combinational* **always** block follows this **reg** declaration. Since this a purely combinational block, it is sensitive to all its inputs, namely x and p_state. Immediately following the block heading, n_state and z are set to their inactive or reset values. This is done so that these variables are always reset with the clock to make sure they do not retain their old values. As discussed before, retaining old values implies latches, which is not what we want in our combinational block.

The body of the combinational **always** block of Figure 3.57 contains a **case**-statement that uses the p_state input of the **always** block for its **case**-expression. This expression is checked against the states of the Moore machine. As in the other styles discussed before, this **case**-statement has **case**-alternatives for *reset, got1, got10,* and *got101* states.

In a block corresponding to a **case**-alternative, based on input values, n_state and z output are assigned values. Unlike the other styles where *current* is used both for the present and next states, here we use two different variables, p_state and n_state.

The next procedural block shown in Figure 3.57 handles the register part of the Huffman model of Figure 3.56. In this part, n_state is treated as the register input and p_state as its output. On the positive edge of the clock, p_state is either set to the *reset* state (**00**) or is loaded with contents of n_state.

Together, *combinational* and *register* blocks describe our state machine in a very modular fashion.

```verilog
module moore_detector ( x, rst, clk, z );
 input x, rst, clk;
 output z;
 reg z;
 parameter [1:0]
  reset = 2'b00, got1 = 2'b01, got10 = 2'b10, got101 = 2'b11;

 reg [1:0] p_state, n_state;

 always @ ( p_state or x ) begin : combinational
  n_state = 0; z = 0;
  case ( p_state )
   reset: begin
      if( x==1'b1 ) n_state = got1;
      else n_state = reset; z = 1'b0;
     end
   got1: begin
      if( x==1'b0 ) n_state = got10;
      else n_state = got1; z = 1'b0;
     end
   got10: begin
      if( x==1'b1 ) n_state = got101;
      else n_state = reset; z = 1'b0;
    end
   got101: begin
      if( x==1'b1 ) n_state = got1;
      else n_state = got10; z = 1'b1;
    end
   default: n_state = reset;
  endcase
 end

 always @( posedge clk ) begin : register
  if( rst ) p_state = reset;
  else p_state = n_state;
 end

endmodule
```

Figure 3.57 Verilog Huffman Coding Style

The advantage of this style of coding is in its modularity and defined tasks of each block. State transitions are handled by the *combinational* block and clocking is done by the *register* block. Changes in clocking, resetting, enabling or presetting the machine only affect the coding of the *register* block. If we were to change the synchronous resetting to asynchronous, the only change we had to make was adding ***posedge*** *rst* to the sensitivity list of the register block.

A More Modular Style. For a design with more input and output lines and more complex output logic, the *combinational* block may further be partitioned into a block for handling transitions and another for assigning values to the outputs of the circuit. For coding both of these blocks, it is necessary to follow the rules discussed for combinational blocks in Section 3.2.4.

```
module mealy_detector ( x, en, clk, rst, z );
  input x, en, clk, rst; output z;   reg z;
  parameter [1:0]  reset = 0, got1 = 1, got10 = 2, got11 = 3;

  reg [1:0] p_state, n_state;

  always @( p_state or x ) begin : Transitions
   n_state = reset;
   case ( p_state )
    reset: if ( x == 1'b1 ) n_state = got1;
      else n_state = reset;
    got1: if ( x == 1'b0 ) n_state = got10;
      else n_state = got11;
    got10: if ( x == 1'b1 ) n_state = got1;
      else n_state = reset;
    got11: if ( x == 1'b1 ) n_state = got11;
      else n_state = got10;
    default:  n_state = reset;
   endcase
  end

  always @( p_state or x )  begin: Outputting
   z = 0;
   case ( p_state )
    reset:  z = 1'b0;
    got1:  z = 1'b0;
    got10:  if ( x == 1'b1 ) z = 1'b1;
      else z = 1'b0;
    got11:  if ( x==1'b1 ) z = 1'b0;
      else z = 1'b1;
    default:  z = 1'b0;
   endcase
  end

  always @ ( posedge clk )  begin: Registering
   if( rst ) p_state = reset;
   else if( en ) p_state = n_state;
  end

endmodule
```

Figure 3.58 Separate Transition and Output Blocks

Figure 3.58 shows the coding of the **110-101** Moore detector using two separate blocks for assigning values to *n_state* and the z output. In a situation like what we have in which the output logic is fairly simple, a simple **assign** statement could replace the *outputting* procedural block. In this case, z must be a **net** and not a **reg**.

The examples discussed above, in particular, the last two styles, show how combinational and sequential coding styles can be combined to describe very complex digital systems.

3.3.5   Memories

Verilog allows description and use of memories. Memories are two-dimensional variables that are declared as **reg**. Verilog only allows **reg** data types for memories. Figure 3.59 shows a **reg** declaration declaring *mem* and its corresponding block diagram. This figure also shows several valid memory operations.

```
reg [7:0] mem [0:1023];
reg [7:0] data;
reg [3:0] short_data;
wire [9:0] addr;
    .
    .
    .
data = mem [956];
    .
    .
    .
short_data = data [7:4];
    .
    .
    .
mem [932] = data;
mem [addr] = {4'b0, short_data};
```

Figure 3.59 Memory Representation

Square brackets that follow the **reg** keyword specify the word-length of the memory. The square brackets that follow the name of the memory (*mem*), specify its address space. A memory can be read by addressing it within its address range, e.g., *mem[956]*. Part of a word in a memory cannot be read directly, i.e., slicing a memory word is not possible. To read part of a word, the whole word must first be read in a variable and then slicing done on this variable. For example, *data[7:4]* can be used after a memory word has been placed into *data*.

With proper indexing, a memory word can be written into by placing the memory name and its index on the left hand side of an assignment, e.g., *mem[932] = data;* , memories can also be indexed by **reg** or **net** type variables, e.g., *mem[addr]*, when *addr* is a 10-bit address bus. Writing into a part of the memory is not possible. In all cases data directly written into a memory word affects all bits of the word being written into. For example to write the four-bit *short_data* into a location of *mem*, we have to decide what goes into the other four bits of the memory word.

Figure 3.60 shows a memory block with separate input and output busses. Writing into the memory is clocked, while reading from it only requires *rw* to be **1**. An **assign** statement handles reading and an **always** block performs writing into this memory.

```
module memory (inbus, outbus, addr, clk, rw);
    input [7:0] inbus;
    input [9:0] addr;
    output [7:0] outbus;
    input clk, rw;

    reg [7:0] mem [0:1023];

    assign outbus = rw ? mem [addr] : 8'bz;

    always @ (posedge clk)
        if (rw == 0) mem [addr] = inbus;

endmodule
```

Figure 3.60 Memory Description

3.4 Writing Testbenches

Verilog coding styles discussed so far were for coding hardware structures, and in all cases synthesizability and direct correspondence to hardware were our main concerns. On the other hand, testbenches do not have to have hardware correspondence and they usually do not follow any synthesizability rules. We will see that delay specifications, and **initial** statements that do not have a one-to-one hardware correspondence are used generously in testbenches.

For demonstration of testbench coding styles, we use the Verilog code of Figure 3.61 that is a **101** Moore detector, as the circuit to be tested.

This description is functionally equivalent to that of Figure 3.50. The difference is in the use of condition expressions (**?:**) instead of **if-else** statements, and separating the output assignment from the main **always** block. This code will be instantiated in the testbenches that follow.

3.4.1 Generating Periodic Data

Figure 3.62 shows a testbench module that instantiates *moore_detector* and applies test data to its inputs. The first statement in this code is the **'timescale** directive that defines the time unit of this description. The testbench itself has no ports, which is typical of all testbenches. All data inputs to a circuit-under-test are locally generated in its testbench.

```
module moore_detector ( x, rst, clk, z );
  input x, rst, clk;
  output z;
  parameter [1:0] a=0, b=1, c=2, d=3;
  reg [1:0] current;

  always @( posedge clk )
    if ( rst )   current = a;
    else case ( current )
      a : current = x ? b : a ;
      b : current = x ? b : c ;
      c : current = x ? d : a ;
      d : current = x ? b : c ;
      default : current = a;
    endcase
  assign z = (current==d) ? 1'b1 : 1'b0;
endmodule
```

Figure 3.61 Circuit Under Test

Because we are using procedural statements for assigning values to ports of the circuit-under-test, all variables mapped with the input ports of this circuit are declared as **reg**. The testbench uses two **initial** blocks and two **always** blocks. The first initial block initializes *clock*, *x*, and *reset* to **0**, **0**, and **1** respectively. The next **initial** block waits for 24 time units (ns in this code), and then sets *reset* back to **0** to allow the state machine to operate.

The **always** blocks shown produce periodic signals with different frequencies on *clock* and *x*. Each block waits for a certain amount of time and then it complements its variable. Complementing begins with the initial values of *clock* and *x* as set in the first **initial** block. We are using different periods for *clock* and *x*, so that a combination of patterns on these circuit inputs is seen. A more deterministic set of values could be set by specifying exact values at specific times.

```
`timescale 1 ns / 100 ps

module test_moore_detector;
  reg x, reset, clock;
  wire z;
  moore_detector uut ( x, reset, clock, z );
  initial begin
    clock=1'b0; x=1'b0; reset=1'b1;
  end
  initial #24 reset=1'b0;
  always #5 clock=~clock;
  always #7 x=~x;
endmodule
```

Figure 3.62 Generating Periodic Data

3.4.2 Random Input Data

Instead of the periodic data on *x* we can use the **$random** predefined system function to generate random data for the *x* input. Figure 3.63 shows such a testbench.

This testbench also combines the two **initial** blocks for initially activating and deactivating *reset* into one. In addition, this testbench has an **initial** block that finishes the simulation after 165 ns.

When the flow into a procedural block reaches the **$finish** system task, the simulation terminates and exits. Another simulation control task that is often used is the **$stop** task that only stops the simulation and allows resumption of the stopped simulation run.

```
`timescale 1 ns / 100 ps

module test_moore_detector;
  reg x, reset, clock;
  wire z;
  moore_detector uut( x, reset, clock, z );
  initial  begin
      clock=1'b0; x=1'b0; reset=1'b1;
      #24 reset=1'b0;
  end
  initial #165 $finish;
  always #5 clock=~clock;
  always #7 x=~x;
endmodule
```

Figure 3.63 Random Data Generation

3.4.3 Synchronized Data

Independent data put into various inputs of a circuit may not be random enough to be able to catch many design errors. Figure 3.64 shows another testbench for our Moore detector that only reads random data into the *x* input after the positive edge of the *clock*.

The third **initial** statement shown in this code uses the **forever** construct to loop forever. Every time when the positive edge of *clock* is detected, after 3 nanoseconds a new random value is put into *x*. The **initial** statement in charge of clock generation uses a **repeat** loop to toggle the *clock* 13 times every 5 nanoseconds and stop. This way, after *clock* stops, all activities cease, and the simulation run terminates. For this testbench we do not need a simulation control task.

The testbench of Figure 3.64 uses an invocation of **$monitor** task to display the contents of the *current* state of the sequence detector every time it changes. The **initial** statement that invokes this task puts it in the background and every time *uut.current* changes, **$monitor** reports its new value. The *uut.current* name is a hierarchical name that uses the instance name of the circuit-under-test to look at its internal variable, *current*.

```
`timescale 1ns/100ps

module test_moore_detector;
  reg x, reset, clock;
  wire z;

  moore_detector uut( x, reset, clock, z );

  initial begin
    clock=1'b0; x=1'b0; reset=1'b1;
    #24 reset=1'b0;
  end
  initial repeat(13) #5 clock=~clock;
  initial forever @(posedge clock) #3 x=$random;
  initial $monitor("New state is %d and occurs at %t", uut.current, $time);
  always @(z) $display("Output changes at %t to %b", $time, z);

endmodule
```

Figure 3.64 Synchronized Test Data

The **$monitor** task shown also reports the time that *current* takes a new value. This time is reported by the **$time** task. The *uut.current* variable uses the decimal format (**%d**) and **$time** is reported using the time format (**%t**). Binary, Octal and Hexadecimal output can be obtained by using **%b**, **%o**, and **%h** format specifications.

The last statement in this testbench is an **always** statement that is sensitive to z. This statement uses the **$display** task to report values put on z. The **$display** task is like the **$monitor**, except that it only becomes active when flow into a procedural block reaches it. When z changes, flow into the **always** statement begins and the **$display** task is invoked to display the new value of z and its time of change. This output is displayed in binary format. Using **$monitor** inside an **initial** statement, for displaying z (similar to that for *uut.current*) would result in exactly the same thing as the **$display** inside an **always** block that is sensitive to z.

3.4.4   Applying Buffered Data

Examples discussed above use random or semi-random data on the x input of the circuit being tested. It is possible that we never succeed in giving x appropriate data to generate a **1** on the z output of our sequence detector. To correct this situation, we define a buffer, put the data we want in it and continuously apply this data to the x input.

Figure 3.65 shows another testbench for our sequence detector of Figure 3.61. In this testbench the 5-bit *buff* variable is initialized to contain **10110**. The **initial** block that follows the clock generation block, rotates concatenation of x and *buff* one place to the right 3 nanoseconds after every time the clock ticks. This process repeats for as long as the circuit clock ticks.

```
`timescale 1ns/100ps

module test_moore_detector;
  reg x, reset, clock;
  wire z;

  reg [4:0] buff;
  initial buff = 5'b10110;

  moore_detector uut( x, reset, clock, z );

  initial begin
    clock=1'b0; x=1'b0; reset=1'b1;
    #24 reset=1'b0;
  end

  initial repeat(18) #5 clock=~clock;
  initial forever @(posedge clock) #3 {buff,x}={x,buff};
  initial forever @(posedge clock) #1 $display(z, uut.current);

endmodule
```

Figure 3.65 Buffered Test Data

The last **initial** statement in this description outputs z and *uut.current* 1 nanosecond after every time the clock ticks. The **$display** task used for this purpose is unformatted which defaults to the decimal data output.

3.4.5 Timed Data

A very simple testbench for our sequence detector can be done by applying test data to x and timing them appropriately to generate the sequence we want, very similar to the way values were applied to *reset* in the previous examples. Figure 3.66 shows this simple testbench.

Techniques discussed in the above examples are just some of what one can do for test data generation. These techniques can be combined for more complex examples. After using Verilog for some time, users form their own test generation techniques. For small designs, simulation environments generally provide waveform editors and other tool-dependent test generation schemes. Some tools come with code fragments that can be used as templates for testbenches.

An important issue is developing testbenches is external file IO. Verilog allows the use of **$readmemh** and **$readmemb** system tasks for reading hex and binary test data into a declared memory. Moreover, for writing responses from a circuit-under-test to an external file, **$fdisplay** can be used. Examples for these features of the language will be shown in Chapters 11 and 14.

```
`timescale 1ns/100ps

module test_moore_detector;
  reg x, reset, clock;
  wire z;

  moore_detector uut( x, reset, clock, z );

  initial begin
    clock=1'b0; x=1'b0; reset=1'b1;
    #24 reset=1'b0;
  end

  always #5 clock=~clock;

  initial begin
    #7  x=1;
    #5  x=0;
    #18 x=1;
    #21 x=0;
    #11 x=1;
    #13 x=0;
    #33 $stop;
  end

endmodule
```

Figure 3.66 Timed Test Data Generation

3.5   Synthesis Issues

Verilog constructs described in this chapter included those for cell modeling as well as those for designs to be synthesized. In describing an existing cell, timing issues are important and must be included in the Verilog code of the cell. At the same time, description of an existing cell may require parts of this cell to be described by interconnection of gates and transistors. On the other hand, a design to be synthesized does not include any timing information because this information is not available until the design is synthesized, and designers usually do not use gates and transistors for high level descriptions for synthesis.

Considering the above, taking timing out of the descriptions, and only using gates when we really have to, the codes presented in this chapter all have one-to-one hardware correspondence and are synthesizable. For synthesis, a designer must consider his or her target library to see what and how certain parts can be synthesized. For example, most FPGAs do not have internal three-state structures and three-state bussings are converted to AND-OR busses.

3.6 Summary

This chapter presented the Verilog HDL language from a hardware design point of view. The chapter used complete design examples at various levels of abstraction for showing ways in which Verilog could be used in a design. We showed how timing details could be incorporated in cell descriptions. Aside from this discussion of timing, all examples that were presented had one-to-one hardware correspondence and were synthesizable. We have shown how combinational and sequential components can be described for synthesis and how a complete system can be put together using combinational and sequential blocks for it to be tested and synthesized.

This chapter did not cover all of Verilog, but only the most often used parts of the language.

4 Programmable Logic Devices

The need for getting designs done quickly has led to the creation and evolution of Programmable Logic devices. The idea began from Read Only Memories (ROM) that were just an organized array of gates and has evolved into System On Programmable Chips (SOPC) that use programmable devices, memories and configurable logic all on one chip.

This chapter shows the evolution of basic array structures like ROMs into complex CPLD (Complex Programmable Logic Devices) and FPGAs (Field Programmable Gate Array). This topic can be viewed from different angles, like logic structure, physical design, programming technology, transistor level, software tools, and perhaps even from historic and commercial aspects. However our treatment of this subject is more at the structural level. We discuss gate level structures of ROMs, PLAs, PALs, CPLDs, and FPGAs. The material is at the level needed for understanding configuration and utilization of CPLDs and FPGAs in digital designs.

4.1 Read Only Memories

We present structure of ROMs by showing the implementation of a 3-input 4-output logic function. The circuit with the truth table shown in Figure 4.1 is to be implemented.

4.1.1 Basic ROM Structure

The simplest way to implement the circuit of Figure 4.1 is to form its minterms using AND gates and then OR the appropriate minterms for formation of the four circuit outputs. The circuit requires eight 3-input AND gates and four OR gates that can take up-to eight inputs. It is easiest to draw this structure in an array format as shown in Figure 4.2.

m:	a	b	c	w	x	y	z
0:	0	0	0	0	0	0	1
1:	0	0	1	1	1	0	0
2:	0	1	0	1	0	1	1
3:	0	1	1	1	0	0	1
4:	1	0	0	0	0	0	1
5:	1	0	1	0	1	1	1
6:	1	1	0	0	0	1	0
7:	1	1	1	0	0	0	0

Figure 4.1 A Simple Combinational Circuit

The circuit shown has an array of AND gates and an array of OR gates, that are referred to as the AND-plane and the OR-plane. In the AND-plane all eight minterms for the three inputs, a, b, and c are generated. The OR plane uses only the minterms that are needed for the outputs of the circuit. See for example minterm 7 that is generated in the AND-plane but not used in the OR-plane. Figure 4.3 shows the block diagram of this array structure.

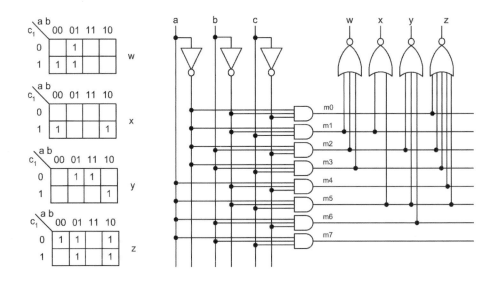

Figure 4.2 AND-OR Implementation

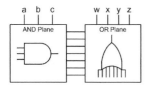

Figure 4.3 AND and OR Planes

4.1.2   NOR Implementation

Since realization of AND and OR gates in most technologies are difficult and generally use more delays and chip area than NAND or NOR implementations, we implement our example circuit using NOR gates. Note that a NOR gate with complemented outputs is equivalent to an OR, and a NOR gate with complemented inputs is equivalent to an AND gate. Our all NOR implementation of Figure 4.4 uses NOR gates for generation of minterms and circuit outputs. To keep functionality and activity levels of inputs and outputs intact, extra inverters are used on the circuit inputs and outputs. These inverters are highlighted in Figure 4.4. Although NOR gates are used, the left plane is still called the AND-plane and the right plane is called the OR-plane.

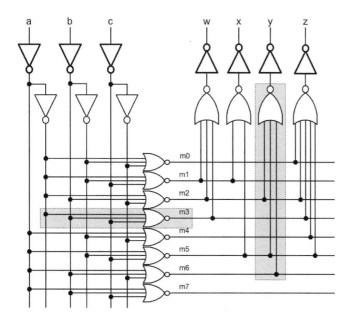

Figure 4.4 All NOR Implementation

4.1.3   Distributed Gates

Hardware implementation of the circuit of Figure 4.4 faces difficulties in routing wires and building gates with large number of inputs. This problem becomes more critical when we are using arrays with tens of inputs. Take for example, a circuit with 16 inputs, which is very usual for combinational circuits. Such a circuit has 64k (2^{16}) minterms. In the AND-plane, wires from circuit inputs must be routed to over 64,000 NOR gates. In the OR-plane, the NOR gates must be large enough for every minterm of the function (over 64,000 minterms) to reach their inputs.

Such an implementation is very slow because of long lines, and takes too much space because of the requirement of large gates. The solution to this problem is to distribute gates along array rows and columns.

In the AND-plane, instead of having a clustered NOR gate for all inputs to reach to, the NOR gate is distributed along the rows of the array. In Figure 4.4, the NOR gate that implements minterm 3 is highlighted. Distributed transistor-level logic of this NOR gate is shown in Figure 4.5. This figure also shows a symbolic representation of this structure.

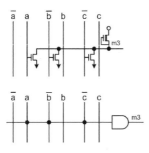

Figure 4.5 Distributed NOR of the AND-plane

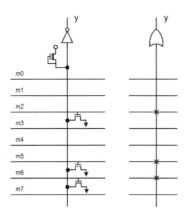

Figure 4.6 Distributed NOR Gate of Output y

Likewise, in the OR-plane, instead of having large NOR gates for the outputs of the circuit, transistors of output NOR gates are distributed along the corresponding output columns. Figure 4.6 shows the distributed NOR structure of the y output of circuit of Figure 4.4. A symbolic representation of this structure is also shown in this figure.

As shown in Figure 4.5 and Figure 4.6, distributed gates are symbolically represented by gates with single inputs. In each case, connections are made on the inputs of the gate. For the AND-plane, the inputs of the AND gate are a, b, and c forming minterm 3, and for the OR gate of Figure 4.6, the inputs of the gate are $m2$, $m5$ and $m6$. The reason for the difference in notations of connections in the AND-plane and the OR-plane (dots versus crosses) becomes clear after the discussion of the next section.

4.1.4 Array Programmability

For the a, b and c inputs, the structure shown in Figure 4.4 implements w, x, y and z functions. In this implementation, independent of our outputs, we have generated all minterms of the three inputs. For any other functions other than w, x, y and z, we would still generate the same minterms, but use them differently. Hence, the AND-plane with which the minterms are generated can be wired independent of the functions realized. On the contrary, the OR-plane can only be known when the output functions have been determined.

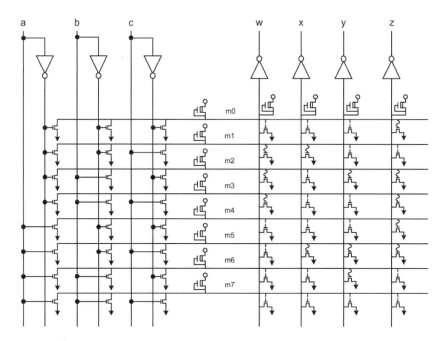

Figure 4.7 Fixed AND-plane, Programmable OR-plane

We can therefore generate a general purpose array logic with all minterms in its AND-plane, and capability of using any number of the minterms for any of the array outputs in its OR-plane. In other words, we want a fixed AND-plane and a programmable (or configurable) OR-plane. As shown in Figure 4.7, transistors for the implementation of minterms in the AND-plane are fixed, but in the OR-plane there are fusible transistors on every output column for every minterm of the AND-plane. For realization of a certain function on an output of this array, transistors corresponding to the used minterms are kept, and the rest are blown to eliminate contribution of the minterm to the output function.

Figure 4.7 shows configuration of the OR-plane for realizing outputs shown in Figure 4.1. Note for example that for output y, only transistors on rows $m2$, $m5$, and $m6$ are connected and the rest are fused off.

Instead of the complex transistor diagram of Figure 4.7, the notation shown in Figure 4.8 is used for representing the programmability of the configurable arrays. The dots in the AND-plane indicate permanent connections, and the crosses in the OR-plane indicate programmable or configurable connections.

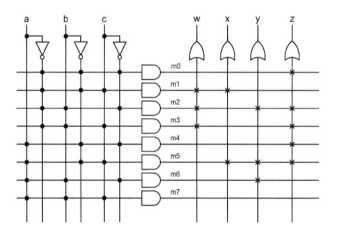

Figure 4.8 Fuse Notation for Configurable Arrays

4.1.5   Memory View

Let us look at the circuit of Figure 4.8 as a black box of three inputs and four outputs. In this circuit, if an input value between 0 and 7 is applied to the abc inputs, a 4-bit value is read on the four circuit outputs. For example $abc=011$ always reads $wxyz=1001$.

If we consider abc as the address inputs and $wxyz$ as the data read from abc designated address, then the black box corresponding to Figure 4.8 can be regarded as a memory with an address space of 8 words and data of four bits wide. In this case, the fixed AND-plane becomes the memory decoder, and the programmable OR-plane becomes the memory array (see Figure 4.9). Because this memory can only be read from and not easily written into, it is referred to as Read Only Memory or ROM.

The basic ROM is a one-time programmable logic array. Other variations of ROMs offer more flexibility in programming, but in all cases they can be read more easily than they can be written into.

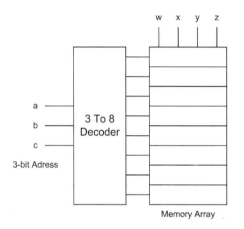

Figure 4.9 Memory View of ROM

4.1.6 ROM Variations

The acronym, ROM is generic and applies to most read only memories. What is today implied by ROM may be ROM, PROM, EPROM, EEPROM or even flash memories. These variations are discussed here.

ROM. ROM is a mask-programmable integrated circuit, and is programmed by a mask in IC manufacturing process. The use of mask-programmable ROMs is only justified when a large volume is needed. The long wait time for manufacturing such circuits makes it a less attractive choice when time-to-market is an issue.

PROM. Programmable ROM is a one-time programmable chip that, once programmed, cannot be erased or altered. In a PROM, all minterms in the AND-plane are generated, and connections of all AND-plane outputs to OR-plane gate inputs are in place. By applying a high voltage, transistors in the OR-plane that correspond to the minterms that are not needed for a certain output are burned out. Referring to Figure 4.7, a fresh PROM has all transistors in its OR-plane connected. When programmed, some will be fused out permanently. Likewise, considering the diagram of Figure 4.8, an un-programmed PROM has X's in all wire crossings in its OR-plane.

EPROM. An Erasable PROM is a PROM that once programmed, can be completely erased and reprogrammed. Transistors in the OR-plane of an EPROM have a normal gate and a floating gate as shown in Figure 4.10. The non-floating gate is a normal NMOS transistor gate, and the floating-gate is surrounded by insulating material that allows an accumulated charge to remain on the gate for a long time.

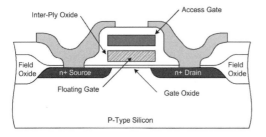

Figure 4.10 Floating Gate

When not programmed, or programmed as a '**1**', the floating gate has no extra charge on it and the transistor is controlled by the non-floating gate (access gate). To fuse-out a transistor, or program a '**0**' into a memory location, a high voltage is applied to the access gate of the transistor which causes accumulation of negative charge in the floating-gate area. This negative charge prevents logic **1** values on the access gate from turning on the transistor. The transistor, therefore, will act as an unconnected transistor for as long as the negative charge remains on its floating-gate.

To erase an EPROM it must be exposed to ultra-violate light for several minutes. In this case, the insulating materials in the floating-gates become conductive and these gates start loosing their negative charge. In this case, all transistors return to their normal mode of operation. This means that all EPROM memory contents become **1**, and ready to be reprogrammed.

Writing data into an EPROM is generally about a 1000 times slower than reading from it. This is while not considering the time needed for erasing the entire EPROM.

EEPROM. An EEPROM is an EPROM that can electrically be erased, and hence the name: Electrically Erasable Programmable ROM. Instead of using ultra-violate to remove the charge on the non-floating gate of an EPROM transistor, a voltage is applied to the opposite end of the transistor gate to remove its accumulated negative charge. An EEPROM can be erased and reprogrammed without having to remove it. This is useful for reconfiguring a design, or saving system configurations. As in EPROMs, EEPROMs are non-volatile memories. This means that they save their internal data while not powered. In order for memories to be electrically erasable, the insulating material surrounding the floating-gate must be much thinner than those of the EPROMS. This makes the number of times EEPROMs can be reprogrammed much less than that of EPROMs and in the order of 10 to 20,000. Writing into a byte of an EEPROM is about 500 times slower than reading from it.

Flash Memory. Flash memories are large EEPROMs that are partitioned into smaller fixed-size blocks that can independently be erased. Internal to a system, flash memories are used for saving system configurations. They are used in digital cameras for storing pictures. As external devices, they are used for temporary storage of data that can be rapidly retrieved.

Various forms of ROM are available in various sizes and packages. The popular 27xxx series EPROMs come in packages that are organized as byte addressable memories. For example, the 27256 EPROM has 256K bits of memory that are arranged into 32K bytes. This package is shown in Figure 4.11.

EPROM - 27256 (32Kb x 8)

		M27256		
V_{PP}	1		28	V_{CC}
A12	2		27	A14
A7	3		26	A13
A6	4		25	A8
A5	5		24	A9
A4	6		23	A11
A3	7		22	$\overline{G}$
A2	8		21	A10
A1	9		20	$\overline{E}$
A0	10		19	Q7
Q0	11		18	Q6
Q1	12		17	Q5
Q2	13		16	Q4
V_{SS}	14		15	Q3

Figure 4.11 27256 EPROM

The 27256 EPROM has a Vpp pin that is used for the supply input during read-only operations and is used for applying programming voltage during the programming phase. The 15 address lines address 256K of 8-bit data that are read on to O7 to O0 outputs. Active low CS and OE are for three-state control of the outputs and are used for cascading EPROMs and/or output bussing.

EPROMs can be cascaded for word length expansion, address space expansion or both. For example, a 1Meg 16-bit word memory can be formed by use of a four by two array of 27256s.

4.2 Programmable Logic Arrays

The price we are paying for the high degree of flexibility of ROMs is the large area occupied by the AND-plane that forms every minterm of the inputs of the ROM. PLAs (Programmable Logic Arrays) constitutes an alternative with less flexibility and less use of silicon. For this discussion we look at ROMs as logic circuits as done in the earlier parts of Section 4.1, and not the memory view of the later parts of this section.

For illustrating the PLA structure, we use the 3-input, 4-output example circuit of Figure 4.1. The AND-OR implementation of this circuit that is shown in Figure 4.2 led to the ROM structure of Figure 4.8, in which minterms generated in the AND-plane are used for function outputs in the OR-plane.

An easy step to reduce the area used by the circuit of Figure 4.8 is to implement only those minterms that are actually used. In this example, since minterm 7 is never used, the last row of the array can be completely eliminated. In large ROM structures, there will be a much larger percentage of unused minterms that can be eliminated. Furthermore, if instead of using minterms, we minimize our output functions and only implement the regained product terms we will be able to save even more rows of the logic array.

Figure 4.12 shows Karnaugh maps for minimization of w, x, y and z outputs of table of Figure 4.1. In this minimization sharing product terms between various outputs is particularly emphasized.

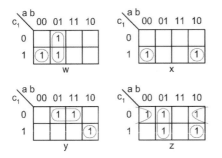

Figure 4.12 Minimizing Circuit of Figure 4.1

Resulting Boolean expressions for the outputs of circuit described by the tables of Figure 4.12 are shown in Figure 4.13. Common product terms in these expressions are vertically aligned.

$w = \bar{a}b + \bar{a}\bar{b}c$

$x = \quad\quad \bar{a}\bar{b}c + a\bar{b}c$

$y = \quad\quad\quad\quad a\bar{b}c + b\bar{c}$

$z = \bar{a}b + \quad\quad a\bar{b}c \quad + b\bar{c}$

Figure 4.13 Minimized Boolean Expressions

Implementation of w, x, y and z functions of a, b and c inputs in an array format using minimized expressions of Figure 4.13 is shown in Figure 4.14. This array uses five rows that correspond to the product terms of the four output functions. Comparing this with Figure 4.8, we can see that we are using less number of rows by generating only the product terms that are needed and not every minterm.

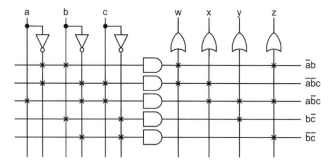

Figure 4.14 PLA Implementation

The price we are paying for the area gained in the PLA implementation of Figure 4.14 is that we now have to program both AND and OR planes. In Figure 4.14 we use X's in both planes where in Figure 4.8 dots are used in the fixed AND-plane and X's in the programmable OR-plane.

While ROM structures are used for general purpose configurable packages, PLAs are mainly used as structured array of hardware for implementing on-chip logic. A ROM is an array logic with fixed AND-plane and programmable OR-plane, and PLA is an array with programmable AND-plane and programmable OR-plane.

A configurable array logic that sits between a PLA and a ROM is one with a programmable AND-plane and a fixed OR-plane. This logic structure was first introduced by Monilitic Memories Inc. (MMI) in the late 1970s and because of its similarity to PLA was retuned to PAL or Programmable Array Logic.

The rationale behind PALs is that outputs of a large logic function generally use a limited number of product terms and the capability of being able to use all product terms for all function outputs is in most cases not utilized. Fixing the number of product terms for the circuit outputs significantly improves the speed of PALs.

4.2.1   PAL Logic Structure

In order to illustrate the logical organization of PALs, we go back to our 3-input, 4-output example of Figure 4.1. Figure 4.15 shows PAL implementation of this circuit. This circuit uses w, x, y and z expressions shown in Figure 4.13. Recall that these expressions are minimal realizations for the outputs of our example circuit and are resulted from the k-maps of Figure 4.12.

The PAL structure of Figure 4.15 has a programmable AND-plane and a fixed OR-plane. Product terms are formed in the AND-plane and three such terms are used as OR gate inputs in the OR-plane. This structure allows a maximum of three product terms per output.

Implementing expressions of Figure 4.13 is done by programming fuses of the AND-plane of the PAL. The z output uses all three available product terms and all other outputs use only two.

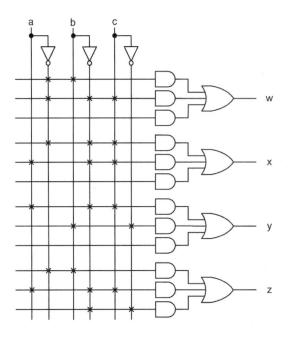

Figure 4.15 PAL Implementation

4.2.2   Product Term Expansion

The limitation on the number of product terms per output in a PAL device can be overcome by providing feedbacks from PAL outputs back into the AND-plane. These feedbacks are used in the AND-plane just like regular inputs of the PAL. Such a feedback allows ORing a subset of product terms of a function to be fed back into the array to further be ORed with the remaining product terms of the function.

Consider for example, PAL implementation of expression w shown below:

$$w = \bar{a}\cdot b \cdot \bar{c} + \bar{a}\cdot \bar{b}\cdot c + a \cdot \bar{b} + a \cdot b \cdot c$$

Let us assume that this function is to be implemented in a 3-input PAL with three product terms per output and with outputs feeding back into the AND-plane, as shown in Figure 4.16.

The partial PAL shown in this figure allows any of its outputs to be used as a circuit primary output or as a partial sum-of-products to be completed by ORing more product terms to it. For implementation of expression w, the first three product terms are generated on the o_1. The structure shown does not allow the last product term $(a \cdot b \cdot c)$ to be ORed on the same output. Therefore, the feedback from this output is used as an input into the next group of product terms. The circled X connection in this figure causes o_1 to be used as an input into the o_2 group. The last product term $(a.b.c)$ is generated in the AND-plane driving the o_2 output and is ORed with o_1 using the OR-gate of the o_2 output. Expression w is generated on o_2. Note that the feedback of o_2 back into the AND-plane does exist, but not utilized.

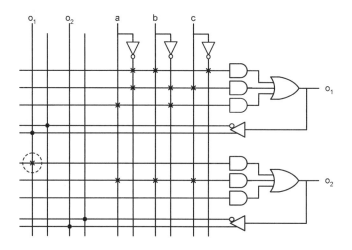

Figure 4.16 A PAL with Product Term Expandability

4.2.3 Three-State Outputs

A further improvement to the original PAL structure of Figure 4.15 is done by adding three-state controls to its outputs as shown in the partial structure of Figure 4.17.

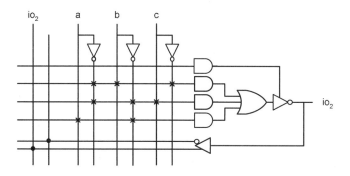

Figure 4.17 PAL Structure with Three Output Control

In addition to the feedback from the output, this structure has two more advantages. First, the pin used as output or partial sum-of-products terms can also be used as input by turning off the three-state gate that drives it. Note that the lines used for feeding back outputs into the AND-plane in Figure 4.16, become connections from the io_2 input into the AND-plane. The second advantage of this structure is that when $io2$ is used as output it becomes a three-state output that is controlled by a programmable product term.

Instead of using a three-state inverting buffer, an XOR gate with three-state output and a fusible input (see Figure 4.18) provides output polarity control when the bi-directional io_2 port is used as output

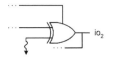

Figure 4.18 Output Inversion Control

4.2.4   Registered Outputs

A major advantage of PALs over PLAs and ROMs is the capability of incorporating registers into the logic structure. Where registers can only be added to the latter two structures on their inputs and outputs, registers added to PAL arrays become more integrated in the input and output of the PAL logic.

As an example structure, consider the registered output of Figure 4.19. The input/output shown can be used as a registered output with three-state, as a two-state output, as a registered feedback into the logic array, or as an input into the AND-plane.

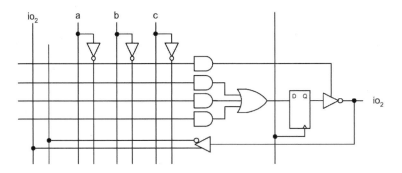

Figure 4.19 Output Inversion Control

A further enhancement to this structure provides logic for bypassing the output flip-flop when its corresponding I/O pin is being used as output. This way, PAL outputs can be programmed as registered or combinational pins.

Other enhancements to the register option include the use of asynchronous control signals for the flip-flop, direct feedback from the flip-flop into the array, and providing a programmable logic function for the flip-flop output and the feedback line.

4.2.5 Commercial Parts

PAL is a trademark of American Micro Devices Inc. More generically, these devices are referred to as PLDs or programmable logic devices. A variation of the original PAL or a PLD that is somewhat different from a PAL is GAL (Generic Array Logic). The inventor of GAL is the Lattice Semiconductor Inc. GALs are electrically erasable; otherwise have a similar logical structure to PALs. By ability to bypass output flip-flops, GALs can be configured as combinational or sequential circuits. To familiarize readers with some actual parts, we discuss one of Altera's PLD devices.

Altera Classic EPLD Family.
Altera Corporation's line of PLDs is its Classic EPLD Family. These devices are EPROM based and have 300 to 900 usable gates depending on the specific part. These parts come in 24 to 68 pin packages and are available in dual in-line package (DIP), plastic J-lead chip carrier (PLCC), pin-grid array (PGA), and small-outline integrated circuit (SOIC) packages. The group of product terms that are ORed together are referred to as a Macrocell, and the number of Macrocells varies between 16 and 48 depending on the device. Each Macrocell has a programmable register that can be programmed as a D, T, JK and SR flip-flop with individual clear and clock controls.

These devices are fabricated on CMOS technology and are TTL compatible. They can be used with other manufacturers PAL and GAL parts. The EP1810 is the largest of these devices that has 900 usable gates, 48 Macrocells, and a maximum of 64 I/O pins. Pin-to-pin logic delay of this part is 20 ns and it can operate with a maximum frequency of 50 MHz. The architecture of this and other Altera's Classic EPLDs includes Macrocells, programmable registers, output enable or clock select, and a feedback select.

Macrocells.
Classic macrocells, shown in Figure 4.20, can be individually configured for both sequential and combinatorial logic operation. Eight product terms form a programmable-AND array that feeds an OR gate for combinatorial logic implementation. An additional product term is used for asynchronous clear control of the internal register; another product term implements either an output enable or a logic-array-generated clock. Inputs to the programmable-AND array come from both the true and complement signals of the dedicated inputs, feedbacks from I/O pins that are configured as inputs, and feedbacks from macrocell outputs. Signals from dedicated inputs are globally routed and can feed the inputs of all device macrocells. The feedback multiplexer controls the routing of feedback signals from macrocells and from I/O pins.

The eight product terms of the programmable-AND array feed the 8-input OR gate, which then feeds one input to an XOR gate. The other input to the XOR gate is connected to a programmable bit that allows the array output to be inverted. This gate is used to implement either active-high or active-low logic, or De Morgan's inversion to reduce the number of product terms needed to implement a function.

Programmable Registers. To implement registered functions, each macrocell register can be individually programmed for D, T, JK, or SR operation. If necessary, the register can be bypassed for combinatorial operation. Registers have an individual asynchronous clear function that is controlled by a dedicated product term. These registers are cleared automatically during power-up. In addition, macrocell registers can be individually clocked by either a global clock or any input or feedback path to the AND array. Altera's proprietary programmable I/O architecture allows the designer to program output and feedback paths for combinatorial or registered operation in both active-high and active-low modes.

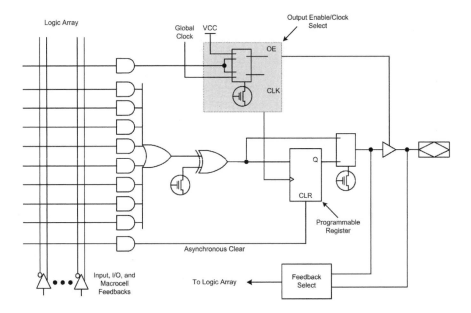

Figure 4.20 Altera's Classic Mecrocell

Output Enable / Clock Select. The box shown in the upper part of Figure 4.20 allows two modes of operations for output and clocking of a Classic macrocell. Figure 4.21 shows these two operating modes (Modes 0 and 1) that are provided by the output enable/clock (OE/CLK) select. The OE/CLK select, which is controlled by a single programmable bit, can be individually configured for each macrocell.

In Mode 0, the tri-state output buffer is controlled by a single product term. If the output enable is high, the output buffer is enabled. If the output enable is low, the output has a high-impedance value. In Mode 0, the macrocell flip-flop is clocked by its global clock input signal.

In Mode 1, the output enable buffer is always enabled, and the macrocell register can be triggered by an array clock signal generated by a product term. This mode allows registers to be individually clocked by any signal on the AND array. With both true and complement signals in the AND array, the register can be configured to trigger on a rising or falling edge. This product-term-controlled clock configuration also supports gated clock structures.

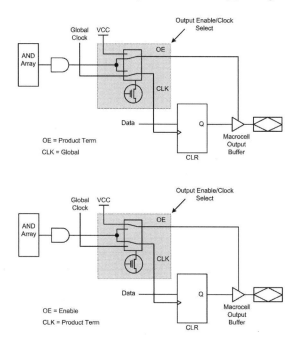

Figure 4.21 Macrocell OE/CLK Select (Upper: Mode 0, Lower: Mode 1)

Feedback Select. Each macrocell in a Classic device provides feedback selection that is controlled by the feedback multiplexer. This feedback selection allows the designer to feed either the macrocell output or the I/O pin input associated with the macrocell back into the AND array. The macrocell output can be either the Q output of the programmable register or the combinatorial output of the macrocell. Different devices have different feedback multiplexer configurations. See Figure 4.22.

EP1810 macrocells can have either of two feedback configurations: quadrant or dual. Most macrocells in EP1810 devices have a quadrant feedback configuration; either the macrocell output or I/O pin input can feed back to other macrocells in the same quadrant. Selected macrocells in EP1810 devices have a dual feedback configuration: the output of the macrocell feeds back to other macrocells in the same quadrant, and the I/O pin input feeds back to all macrocells in the device. If the associated I/O pin is not used, the macrocell output can optionally feed all macrocells in the device. In this case, the output of the macrocell passes through the tri-state buffer and uses the feedback path between the buffer and the I/O pin.

Figure 4.22 Classic Feedback Multiplexer Configurations

Altera "Classic EPLD Family" datasheet describes other features of EP1810 and other Altera's EPLDs. This document has an explanation of device timings of these EPLDs.

4.3   Complex Programmable Logic Devices

The next step up in the evolution and complexity of programmable devices is the CPLD, or Complex PLD. Extending PLDs by making their AND-plane larger and having more macrocells in order to be able to implement larger and more complex logic circuits would face difficulties in speed and chip area utilization. Therefore, instead of simply making these structures larger, CPLDs are created that consist of multiple PLDs with programmable wiring channels between the PLDs. Figure 4.23 shows the general block diagram of a CPLD.

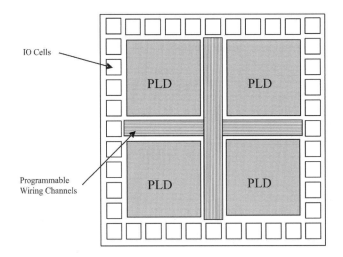

Figure 4.23 CPLD Block Diagram

The approach taken by different manufacturers for implementation of their CPLDs are different. As a typical CPLD we discuss Altera's EPM7128S that is a member of this manufacturer's MAX 7000 Programmable Device Family.

4.3.1   Altera's MAX 7000S CPLD

A member of Altera's MAX 7000 S-series is the EPM7128S CPLD. This is an EEPROM-based programmable logic device with in-system programmability feature through its JTAG interface. Logic densities for the MAX family of CPLDs range from 600 to 5,000 usable gates and the EPM7128S is a mid-rage CPLD in this family with 2,500 usable gates. Note that these figures are 2 to 4 times larger than those of the PLDs from Altera.

The EPM7128s is available in plastic J-lead chip carrier (PLCC), ceramic pin-grid array (PGA), plastic quad flat pack (PQFP), power quad flat pack (RQFP), and 1.0-mm thin quad flat pack (TQFP) packages. The maximum frequency of operation of this part is 147.1 MHz, and it has a propagation delay of 6 ns. This part can operate with 3.3 V or 5.0 V.

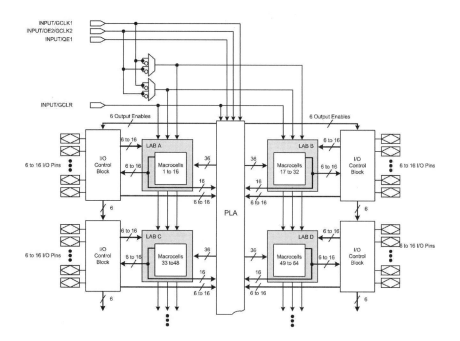

Figure 4.24 Altera's CPLD Architecture

This CPLD has 8 PLDs that are referred to as Logic Array Blocks (LABs). Each LAB has 16 macrocells, making the total number of its macrocells 128. The LABs are linked by a wiring channel that is referred to as the Programmable Interconnect Array (PIA). The macrocells include hardware for expanding product terms by linking several macrocells. The overall architecture of this part is shown in Figure 4.24. In what follows, blocks shown in this figure will be briefly described.

Logic Array Blocks. The EPM7128S has 8 LABs (4 shown in Figure 4.24) that are linked by the PIA global wiring channel. In general, a LAB has the same structure as a PLD described in the previous section. Multiple LABs are linked together via the PIA global bus that is fed by all dedicated inputs, I/O pins, and macrocells. Signals included in a LAB are 36 signals from the PIA that are used for general logic inputs, global controls that are used for secondary register functions, and direct input paths from I/O pins to the registers.

Macrocells. The MAX 7000 macrocell can be individually configured for either sequential or combinatorial logic operation. The macrocell consists of three functional blocks: the logic array, the product-term select matrix, and the programmable register. The macrocell for EPM7128S is shown in Figure 4.25.

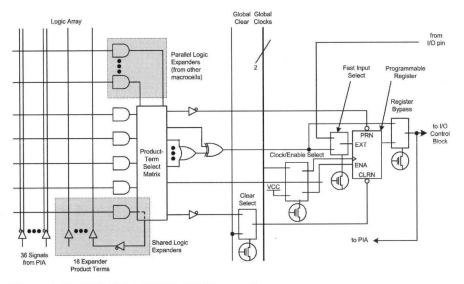

Figure 4.25 MAX 7000 EPM7128S Macrocell

Combinatorial logic is implemented in the logic array, which provides five product terms per macrocell. The product-term select matrix allocates these product terms for use as either primary logic inputs (to the OR and XOR gates) to implement combinatorial functions, or as secondary inputs to the macrocell's register clear, preset, clock, and clock enable control functions. Two kinds of expander product terms ("expanders") are available to supplement macrocell logic resources: Shareable expanders, which are inverted product terms that are fed back into the logic array, and Parallel expanders, which are product terms borrowed from adjacent macrocells.

For registered functions, each macrocell flip-flop can be individually programmed to implement D, T, JK, or SR operation with programmable clock control. The flip-flop can be bypassed for combinatorial operation. Each programmable register can be clocked by a global clock signal and enabled by an active-high clock enable, and by an array clock implemented with a product term. Each register also supports asynchronous preset and clear functions. As shown in Figure 4.25, the product-term select matrix allocates product terms to control these operations.

Expander Product Terms. Although most logic functions can be implemented with the five product terms available in each macrocell, the more complex logic functions require additional product terms. Another macrocell can be used to supply the required logic resources; however, the MAX 7000 architecture also allows both shareable and parallel expander product terms ("expanders") that provide additional product terms directly to any macrocell in the same LAB.

Each LAB has 16 shareable expanders that can be viewed as a pool of uncommitted single product terms (one from each macrocell) with inverted outputs that feed back into the logic array. Each shareable expander can be used and shared by any or all macrocells in the LAB to build complex logic functions. Figure 4.26 shows how shareable expanders can feed multiple macrocells.

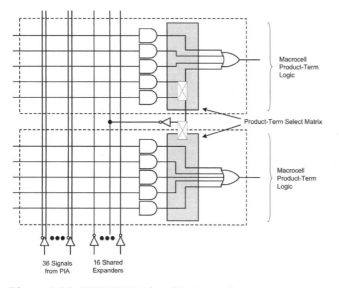

Figure 4.26 MAX 7000 Sharable Expanders

Parallel expanders are unused product terms that can be allocated to a neighboring macrocell. Parallel expanders allow up to 20 product terms to directly feed the macrocell OR logic, with five product terms provided by the macrocell and 15 parallel expanders provided by neighboring macrocells in the LAB.

Two groups of 8 macrocells within each LAB (e.g., macrocells 1 through 8 and 9 through 16) form two chains to lend or borrow parallel expanders. A macrocell borrows parallel expanders from lower numbered macrocells. For example, Macrocell 8 can borrow parallel expanders from Macrocell 7, from Macrocells 7 and 6, or from Macrocells 7, 6, and 5. Within each group of 8, the lowest-numbered macrocell can only lend parallel expanders and the highest-

numbered macrocell can only borrow them. Figure 4.27 shows how parallel
expanders can be borrowed from a neighboring macrocell.

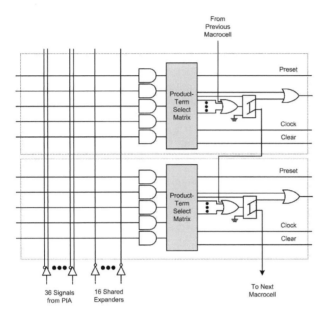

Figure 4.27 MAX 7000 Parallel Expanders

Programmable Interconnect Array.   Logic is routed between LABs via the
programmable interconnect array (PIA).   This global bus is a programmable
path that connects any signal source to any destination on the device. All MAX
7000 dedicated inputs, I/O pins, and macrocell outputs feed the PIA, which
makes the signals available throughout the entire device.   Only the signals
required by each LAB are actually routed from the PIA into the LAB.

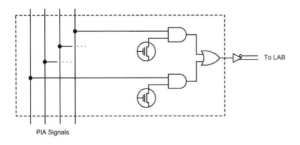

Figure 4.28 PIA Routing in MAX 7000 Devices

Figure 4.28   shows how the PIA signals are routed into the LAB.   An
EEPROM cell controls one input to a 2-input AND gate, which selects a PIA

signal to drive into the LAB. The PIA has a fixed delay that eliminates skew between signals and makes timing performance easy to predict.

I/O Control Blocks. The I/O control block allows each I/O pin to be individually configured for input, output, or bidirectional operation. All I/O pins have a tri-state buffer that is individually controlled by one of the global output enable signals or directly connected to ground or VCC. Figure 4.29 shows the I/O control block for the EPM7128S of the MAX 7000 family. The I/O control block shown here has six global output enable signals that are driven by the true or complement of two output enable signals, a subset of the I/O pins, or a subset of the I/O macrocells.

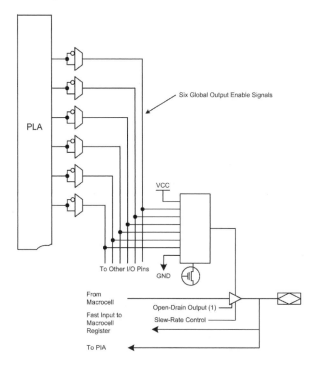

Figure 4.29 I/O Control Block for EPM7128S

When the tri-state buffer control is connected to ground, the output is tri-stated (high impedance) and the I/O pin can be used as a dedicated input. When the tri-state buffer control is connected to VCC, the output is enabled. The MAX 7000 architecture provides dual I/O feedback, in which macrocell and pin feedbacks are independent. When an I/O pin is configured as an input, the associated macrocell can be used for buried logic.

Because of the logic nature of this book, the above discussion concentrated on the logical architecture of the EPM7128S member of the MAX 7000S family. Other details of this part including its timing parameters, programming alternatives, and its In-System Programmability (ISP) features can be found in

Altera's "*MAX 7000 Programmable Logic Device Family*" datasheet. In addition to the EPM7128S device that we discussed, this datasheet has details about other members of the MAX 7000 CPLD family.

4.4   Field Programmable Gate Arrays

A more advanced programmable logic than the CPLD is the Field Programmable Gate Array (FPGA). An FPGA is more flexible than CPLD, allows more complex logic implementations, and can be used for implementation of digital circuits that use equivalent of several Million logic gates.

An FPGA is like a CPLD except that its logic blocks that are linked by wiring channels are much smaller than those of a CPLD and there are far more such logic blocks than there are in a CPLD. FPGA logic blocks consist of smaller logic elements. A logic element has only one flip-flop that is individually configured and controlled. Logic complexity of a logic element is only about 10 to 20 equivalent gates. A further enhancement in the structure of FPGAs is the addition of memory blocks that can be configured as a general purpose RAM. Figure 4.30 shows the general structure of an FPGA.

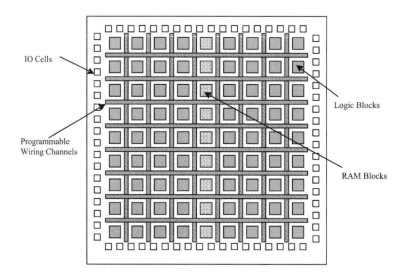

Figure 4.30 FPGA General Structure

As shown in Figure 4.30, an FPGA is an array of many logic blocks that are linked by horizontal and vertical wiring channels. FPGA RAM blocks can also be used for logic implementation or they can be configured to form memories of various word sizes and address space. Linking of logic blocks with the I/O cells and with the memories are done through wiring channels. Within logic blocks, smaller logic elements are linked by local wires.

FPGAs from different manufacturers vary in routing mechanisms, logic blocks, memories and I/O pin capabilities. As a typical FPGA, we will discuss Altera's EPF10K70 that is a member of this manufacturer's FLEX 10K Embedded Programmable Logic Device Family.

4.4.1 Altera's FLEX 10K FPGA

A member of Altera's FLEX 10K family is the EPF10K70 FPGA. This is a SRAM-based FPGA that can be programmed through its JTAG interface. This interface can also be used for FPGAs logic boundary-scan test. Typical gates of this family of FPGAs range from 10,000 to 250,000. This family has up to 40,960 RAM bits that can be used without reducing logic capacity.

Altera's FLEX 10K devices are based on reconfigurable CMOS SRAM elements, the Flexible Logic Element MatriX (FLEX) architecture is geared for implementation of common gate array functions. These devices are reconfigurable and can be configured on the board for the specific functionality required. At system power-up, they are configured with data stored in an Altera serial configuration device or provided by a system controller. Altera offers the EPC1, EPC2, EPC16, and EPC1441 configuration devices, which configure FLEX 10K devices via a serial data stream. Configuration data can also be downloaded from system RAM or from Altera's BitBlaster™ serial download cable or ByteBlasterMV™ parallel port download cable. After a FLEX 10K device has been configured, it can be reconfigured in-circuit by resetting the device and loading new data. Reconfiguration requires less than 320 ms. FLEX 10K devices contain an interface that permits microprocessors to configure FLEX 10K devices serially or in parallel, and synchronously or asynchronously. The interface also enables microprocessors to treat a FLEX 10K device as memory and configure the device by writing to a virtual memory location.

The EPF10K70 has a total of 70,000 typical gates that include logic and RAM. There are a total of 118,000 system gates. The entire array contains 468 Logic Array Blocks (LABs) that are arranged in 52 columns and 9 rows. The LABs are the "Logic Blocks" shown in Figure 4.30. Each LAB has 8 Logic Elements (LEs), making the total number of its LEs 3,744. In the middle of the FPGA chip, a column of 9 Embedded Array Blocks (EABs), each of which has 2,048 bits, form the 18,432 RAM bits of this FPGA. The EPF10K70 has 358 user I/O pins.

FLEX 10K Blocks. The block diagram of a FLEX 10K is shown in Figure 4.31. Each group of LEs is combined into an LAB; LABs are arranged into rows and columns. Each row also contains a single EAB. The LABs and EABs are interconnected by the FastTrack Interconnect. IOEs are located at the end of each row and column of the FastTrack Interconnect.

FLEX 10K devices provide six dedicated inputs that drive the flip-flops' control inputs to ensure the efficient distribution of high-speed, low-skew (less than 1.5 ns) control signals. These signals use dedicated routing channels that provide shorter delays and lower skews than the FastTrack Interconnect. Four of the dedicated inputs drive four global signals. These four global signals can also be driven by internal logic, providing an ideal solution for a clock divider or an internally generated asynchronous clear signal that clears many registers in the device.

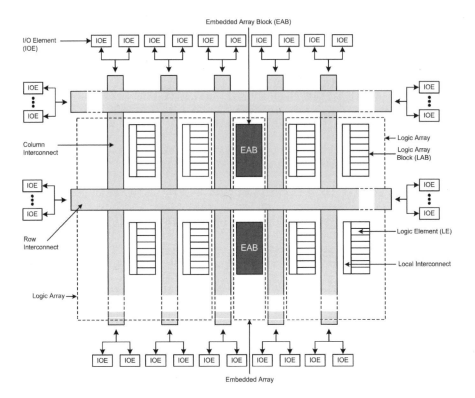

Figure 4.31 FLEX 10K Block Diagram

Signal interconnections within FLEX 10K devices and to and from device pins are provided by the FastTrack Interconnect, a series of fast, continuous row and column channels that run the entire length and width of the device.

Each I/O pin is fed by an I/O element (IOE) located at the end of each row and column of the FastTrack Interconnect. Each IOE contains a bidirectional I/O buffer and a flip-flop that can be used as either an output or input register to feed input, output, or bidirectional signals. When used with a dedicated clock pin, these registers provide exceptional performance. As inputs, they provide setup times as low as 1.6 ns and hold times of 0 ns; as outputs, these registers provide clock-to-output times as low as 5.3 ns. IOEs provide a variety of features, such as JTAG BST support, slew-rate control, tri-state buffers, and open-drain outputs.

Embedded Array Block. Each device contains an embedded array to implement memory and specialized logic functions, and a logic array to implement general logic. The embedded array consists of a series of EABs (EPF10K70 has 9 EABs). When implementing memory functions, each EAB provides 2,048 bits, which can be used to create RAM, ROM, dual-port RAM, or first-in first-out (FIFO) functions. When implementing logic, each EAB can contribute 100 to

600 gates towards complex logic functions, such as multipliers, microcontrollers, state machines, and DSP functions. EABs can be used independently, or multiple EABs can be combined to implement larger functions. Figure 4.32 shows the architecture of EABs and their interconnect busses. The EPF10K70 has 26 inputs to the LAB local interconnect channel from the row.

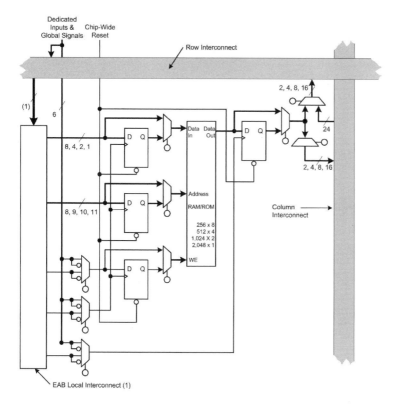

Figure 4.32 EAB Architecture and its Interconnects

Logic functions are implemented by programming the EAB with a read-only pattern during configuration, creating a large look-up table. With tables, combinatorial functions are implemented by looking up the results, rather than by computing them. This implementation of combinatorial functions can be faster than using algorithms implemented in general logic, a performance advantage that is further enhanced by the fast access times of EABs. The large capacity of EABs enables designers to implement complex functions in one logic level. For example, a single EAB can implement a 4×4 multiplier with eight inputs and eight outputs.

EABs can be used to implement synchronous RAM that generates its own WE signal and is self-timed with respect to the global clock. A circuit using the EAB's self-timed RAM need only meet the setup and hold time specifications of the global clock. When used as RAM, each EAB can be configured in any of the following sizes: 256×8, 512×4, 1,024×2, or 2,048×1. Larger blocks of RAM are created by combining multiple EABs.

Different clocks can be used for the EAB inputs and outputs. Registers can be independently inserted on the data input, EAB output, or the address and WE inputs. The global signals and the EAB local interconnect can drive the WE signal. The global signals, dedicated clock pins, and EAB local interconnect can drive the EAB clock signals. Because the LEs drive the EAB local interconnect, the LEs can control the WE signal or the EAB clock signals.

Each EAB is fed by a row interconnect and can drive out to row and column interconnects. Each EAB output can drive up to two row channels and up to two column channels; the unused row channel can be driven by other LEs.

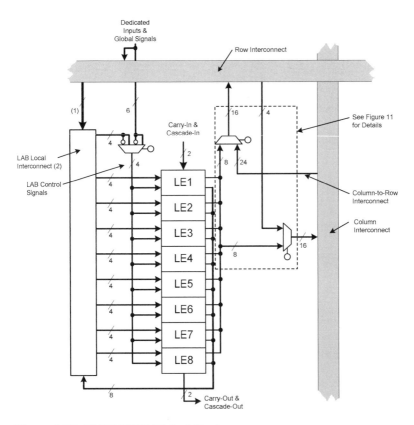

Figure 4.33 FLEX 10K LAB Architecture

Logic Array Block. Referring to Figure 4.31, the logic array of FLEX 10K consists of logic array blocks (LABs). Each LAB contains eight LEs and a local interconnect. An LE consists of a 4-input look-up table (LUT), a programmable flip-flop, and dedicated signal paths for carry and cascade functions. Each LAB represents about 96 usable gates of logic.

Each LAB (see Figure 4.33) provides four control signals with programmable inversion that can be used in all eight LEs. Two of these signals can be used as clocks; the other two can be used for clear/preset control. The LAB clocks can be driven by the dedicated clock input pins, global signals, I/O signals, or internal signals via the LAB local interconnect. The LAB preset and clear control signals can be driven by the global signals, I/O signals, or internal signals via the LAB local interconnect. The global control signals are typically used for global clock, clear, or preset signals because they provide asynchronous control with very low skew across the device. If logic is required on a control signal, it can be generated in one or more LEs in any LAB and driven into the local interconnect of the target LAB. In addition, the global control signals can be generated from LE outputs.

Logic Element. The LE is the smallest unit of logic in the FLEX 10K architecture. Each LE contains a four-input LUT, which is a function generator that can compute any function of four variables. In addition, each LE contains a programmable flip-flop with a synchronous enable, a carry chain, and a cascade chain. Each LE drives both the local and the FastTrack Interconnect. See Figure 4.34.

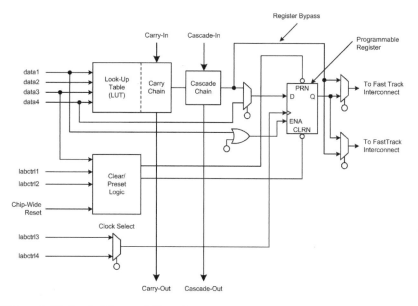

Figure 4.34 Logic Element Structure

The programmable flip-flop in the LE can be configured for D, T, JK, or SR operation. The clock, clear, and preset control signals on the flip-flop can be driven by global signals, general-purpose I/O pins, or any internal logic. For combinatorial functions, the flip-flop is bypassed and the output of the LUT drives the output of the LE.

The LE has two outputs that drive the interconnect; one drives the local interconnect and the other drives either the row or column FastTrack Interconnect. The two outputs can be controlled independently. For example, the LUT can drive one output while the register drives the other output. This feature, called register packing, can improve LE utilization because the register and the LUT can be used for unrelated functions.

The FLEX 10K architecture provides two types of dedicated high-speed data paths that connect adjacent LEs without using local interconnect paths: carry chains and cascade chains. The carry chain supports high-speed counters and adders; the cascade chain implements wide-input functions with minimum delay. Carry and cascade chains connect all LEs in an LAB and all LABs in the same row. Intensive use of carry and cascade chains can reduce routing flexibility. Therefore, the use of these chains should be limited to speed-critical portions of a design.

The FLEX 10K LE can operate in the following four modes: Normal mode, Arithmetic mode, Up/down counter mode, and Clearable counter mode. Each of these modes uses LE resources differently. In each mode, seven available inputs to the LE—the four data inputs from the LAB local interconnect, the feedback from the programmable register, and the carry-in and cascade-in from the previous LE—are directed to different destinations to implement the desired logic function. Three inputs to the LE provide clock, clear, and preset control for the register. The architecture provides a synchronous clock enable to the register in all four modes.

The FLEX 10K architecture, shown in Figure 4.34, includes a "Clear/Preset Logic" block, which provides controls for the LE flip-flop. Logic for the programmable register's clear and preset functions is controlled by the DATA3, LABCTRL1, and LABCTRL2 inputs to the LE. The clear and preset control structure of the LE asynchronously loads signals into a register. Either LABCTRL1 or LABCTRL2 can control the asynchronous clear. Alternatively, the register can be set up so that LABCTRL1 implements an asynchronous load. The data to be loaded is driven to DATA3; when LABCTRL1 is asserted, DATA3 is loaded into the register.

FastTrack Interconnect. In the FLEX 10K architecture, connections between LEs and device I/O pins are provided by the FastTrack Interconnect, shown in Figure 4.35. This is a series of continuous horizontal and vertical routing channels that traverse the device. This global routing structure provides predictable performance, even in complex designs.

The FastTrack Interconnect consists of row and column interconnect channels that span the entire device. Each row of LABs is served by a dedicated row interconnect. The row interconnect can drive I/O pins and feed other LABs in the device. The column interconnect routes signals between rows and can drive I/O pins.

A row channel can be driven by an LE or by one of three column channels. These four signals feed dual 4-to-1 multiplexers that connect to two specific row channels. These multiplexers, which are connected to each LE, allow column channels to drive row channels even when all eight LEs in an LAB drive the row interconnect.

Each column of LABs is served by a dedicated column interconnect. The column interconnect can then drive I/O pins or another row's interconnect to route the signals to other LABs in the device. A signal from the column interconnect, which can be either the output of an LE or an input from an I/O pin, must be routed to the row interconnect before it can enter an LAB or EAB. Each row channel that is driven by an IOE or EAB can drive one specific column channel.

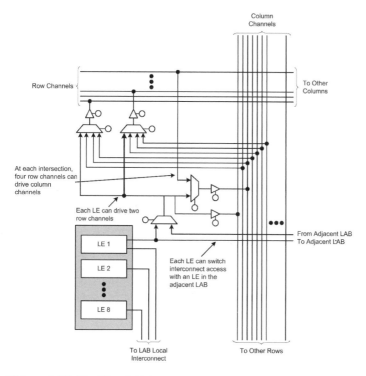

Figure 4.35 FastTrack Interconnect

Access to row and column channels can be switched between LEs in adjacent pairs of LABs. For example, an LE in one LAB can drive the row and column channels normally driven by a particular LE in the adjacent LAB in the same row, and vice versa. This routing flexibility enables routing resources to be used more efficiently. EPF10K70 has 8 rows, 312 channels per row, 52 columns, and 24 interconnects per column.

I/O Element. An I/O element (IOE) of FLEX 10K (see the top-level architecture of Figure 4.31) contains a bidirectional I/O buffer and a register that can be used either as an input register for external data that requires a fast setup time, or as an output register for data that requires fast clock-to-output performance. In some cases, using an LE register for an input register will result in a faster setup time than using an IOE register. IOEs can be used as input, output, or bidirectional pins. For bidirectional registered I/O implementation, the output register should be in the IOE, and the data input and output enable register should be LE registers placed adjacent to the bidirectional pin.

When an IOE connected to a row (as shown in Figure 4.36), is used as an input signal it can drive two separate row channels. The signal is accessible by all LEs within that row. When such an IOE is used as an output, the signal is driven by a multiplexer that selects a signal from the row channels. Up to eight IOEs connect to each side of each row channel.

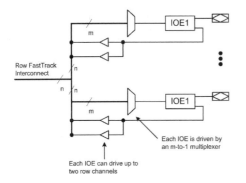

Figure 4.36 FLEX 10K Row-to-IOE Connections

When an IOE connected to a column (as shown in Figure 4.37) is used as an input, it can drive up to two separate column channels. When an IOE is used as an output, the signal is driven by a multiplexer that selects a signal from the column channels. Two IOEs connect to each side of the column channels. Each IOE can be driven by column channels via a multiplexer. The set of column channels that each IOE can access is different for each IOE.

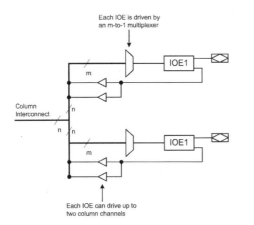

Figure 4.37 FLEX 10K Column-to-IOE Connections

In this section we have shown FPGA structures by using Altera's EPF10K70 that is a member of the FLEX 10K family as an example. The focus of the above discussion was on the logic structure on this programmable device, and many of the timing and logical configuration details have been eliminated. The *"FLEX 10K Embedded Programmable Logic Device Family"* datasheet is a detailed document about this and other FLEX 10K members. Interested readers are encouraged to study this document for advanced features and details of logical configurations of this FPGA family.

4.5 Summary

In an evolutionary fashion, this chapter showed how a simple idea like the ROM have evolved into FPGA programmable chips that can be used for implementation of complete systems that include several processors, memories and even some analog parts. The first part of this chapter discussed generic structures of programmable devices, and in the second part, when describing more complex programmable devices, Altera devices were used as examples. We focused on the structures and tried to avoid very specific manufacturer's details. This introduction familiarizes readers with the general concepts of the programmable devices and enables them to better understand specific manufacturer's datasheets.

5 Computer Architecture

This chapter provides the basic concepts and techniques that are necessary to understand and design a computer system. The chapter begins with an introduction of computer systems, which describes the role of software and hardware in a computer system. In this part, instructions, programs, instruction execution, and processing hardware will be described. After this brief introduction, we will describe a computer from its software point of view. This will be brief, and mainly focuses on definition of terms and the necessary background for understanding the hardware of a processor. In description of the hardware of a computer we use a simple example that has the basic properties found in most processing units. This becomes the main focus of this chapter. In a top-down fashion, we will show control-data partitioning of this example and design and implement the individual parts of this machine.

5.1 Computer System

It is important to understand what it is that we refer to as a computer. This section gives this overall view. A computer is an electronic machine which performs some computations. To have this machine perform a task, the task must be broken into small instructions, and the computer will be able to perform the complete task by executing each of its comprising instructions. In a way, a computer is like any of us trying to evaluate something based on a given algorithm.

To perform a task, we come up with an algorithm for it. Then we break down the algorithm into a set of small instructions, called a program, and using these step-by-step instructions we achieve the given task. A computer does exactly the same thing except that it cannot decide on the algorithm for performing a task, and it cannot break down a task into small instructions either. To use a computer, we come up with a set of instructions for it to do, and it will be able to do these instructions much faster than we could.

Putting ourselves in place of a computer, if we were given a set of instructions (a program) to perform, we would need an instruction sheet and a data sheet (or scratch paper). The instruction sheet would list all instructions to perform. The data sheet, on the other hand, would initially contain the initial data used by the program, and it could also be used for us to write our intermediate and perhaps the final results of the program we were performing.

In this scenario, we read an instruction from the instruction sheet, read its corresponding data from the data sheet, use our brain to perform the instruction, and write the result in the data sheet. Once an instruction is complete, we go on to the next instruction and perform that. In some cases, based on the results obtained, we might skip a few instructions and jump to the beginning of a new set of instructions. We continue execution of the given set of instructions until we reach the end of the program.

For example consider an algorithm that is used to add two 3-digit decimal numbers. Figure 5.1 shows the addition algorithm and two decimal numbers that are added by this algorithm. The algorithm starts with reading the first two digits from the paper, and continues with adding them in the brain and writing the sum and output carry on the paper in their specified positions. So the paper is used to store both the result (i.e., sum), and the temporary results (i.e., carry). The algorithm continues until it reaches Step 7 of Figure 5.1.

There are similar components in a computing machine (computer). A CPU has a memory unit. The part of the memory that is used to store instructions corresponds to the instruction sheet and the part that stores temporary and final results corresponds to the scratch paper or the data sheet. The *Central Processing Unit* (*CPU*), which corresponds to the brain, sequences and executes the instructions.

1. Set column index, i, to 1
2.   Add digits in the ith position
3.   Generate *Sum*, write in result-position i
4.   Generate *Carry*, write in carry-position $i+1$
5.   Increment i
6.   Are there more digits in position i?
     a. If so, jump to *Step 2*
     b. Otherwise, continue
7. End.
	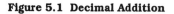

Figure 5.1 Decimal Addition

There is an important difference between storing information in these two methods. In the manual computation the instructions are represented using natural language or some human readable guidelines and data is usually presented in decimal forms. On the other hand, in the computer, information (both instructions and data) are stored and processed in the binary form. To provide communication between the user and the computer, an *input-output* (*IO*) *device* is needed to convert information from human language to machine language (0 and 1) and vice versa. So each computer should have a CPU to execute instructions, a memory to store instructions and data, and an IO device to transfer information between the computer and the outside word.

There are several ways to interconnect these three components (Memory, CPU and IO) in a computer system. A computer with an interconnection shown in Figure 5.2 is called a *Von-Neumann* computer. The CPU communicates with the IO device(s) for receiving input data and displaying results. It communicates with the memory for reading instructions and data as well as writing data.

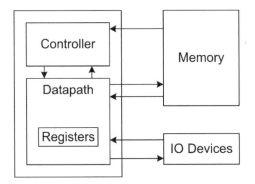

Figure 5.2 Von-Neumann Machine

As shown in this figure, the CPU is divided into *datapath* and *controller*. The datapath has storage elements (registers) to store intermediate data, handles transfer of data between its storage components, and performs arithmetic or logical operations on data that it stores. Datapath also has communication lines for transfer of data; these lines are referred to as busses. Activities in the datapath include reading from and writing into data registers, controlling busses selection of their sources and destinations, and control of the logic units for performing various operations that they are built for.

The controller commands the data-path to perform proper operation(s) according to the instruction it is executing. Control signals carry these commands from the controller to the datapath. Control signals are generated by controller state machine that, at all times, knows the status of the task that is being executed and the sort of the information that is stored in datapath registers. Controller is the thinking part of a CPU.

5.2 Computer Software

The part of a computer system that contains instructions for the machine to perform is called its software. For making software of a computer available for its hardware to execute, it is put in the memory of the computer system. As shown in Figure 5.2 the memory of a system is directly accessible by its hardware. There are several ways computer software can be described.

5.2.1   Machine Language

Computers are designed to perform our commands. To command a computer, you should know the computer alphabets. As mentioned in Chapter 2, the computer alphabet is just two letters **0** and **1**. An individual command, which is presented using these two letters, is called *instruction*. Instructions are binary numbers that are meaningful for a computer. For example, the binary number *0000000010000001* commands a computer to add two numbers. This number is divided into three fields, the first filed (*0000*) signifies the *ADD* operation, and the other two (*000010* and *000001*) are references to the numbers that are to be added. The binary language, in which instructions are defined, is referred to as *machine language*. Note that the machine language is hardware dependent, which means that different computers have different machine languages.

5.2.2   Assembly Language

The earliest programmers wrote their programs in machine language. Machine language programs are tedious and error prone to write, and difficult to understand. So the programmers used a symbolic notation which is closer to the human language. This symbolic language is called *assembly language* which is easier to use than machine language. For example, the above instruction might be written as *ADD A, B*.

We need a special program, called *assembler*, to convert a program from assembly language to machine language. Because the assembly language is a symbolic representation of the machine language, each computer has its own assembler.

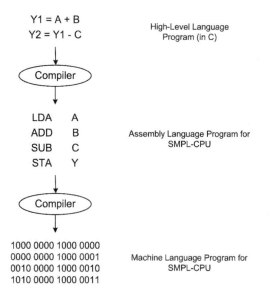

Figure 5.3 Translation of Programming Languages

5.2.3   High-Level Language

The development of the programming language continued and resulted in a *high-level programming language* which is closer to the programmer's problem specification. For example a user can write the *ADD A, B instruction* mentioned above, as *A + B*, which is more readable.

Similarly we need a program, called compiler, to translate these high-level programs into the assembly language of a specific computer. So the high-level languages can be used on different computers.   Figure 5.3 shows the translation of different programming languages. The SMPL CPU shown in this figure is a small fictitious processor that we will use later in this chapter.

5.2.4   Instruction Set Architecture

As shown in Figure 5.4, a computer system is comprised of two major parts, *hardware* and *software*. The interface between these parts is called *Instruction Set Architecture (ISA)*.   ISA defines how data that is being read from a CPU memory (CPU program) and that is regarded as an instruction, is interpreted by the hardware of the CPU.

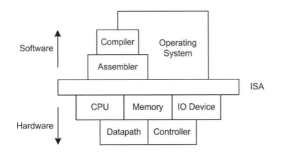

Figure 5.4 Computer System Components

Hardware. The hardware part of a computer has three major components (CPU, Memory Unit, and IO Device). Breaking down the CPU into its composing parts, shows that the CPU is built from an interconnection of datapath and controller.

Datapath consists of *functional units* and *storage elements*. A functional unit (such as an adder, a subtracter and an arithmetic-logical unit (ALU)) performs an arithmetic or logical operation. The storage element (such as a register or a register-file) is needed to store data. *Bussing structure* describes the way functional units and storage elements are connected. Datapath shows the flow of data in the CPU, i.e., how data is stored in the registers and processed by functional units. Controller is used to control the flow of data or the way data is processed in the data-path.

Figure 5.5 shows a datapath which contains two registers R0, and R1 and an adder/subtracter unit. This datapath is able to add (subtract) the content of R0 to (from) R1 and store the result in R0. The controller controls the data-path to perform addition or subtraction.

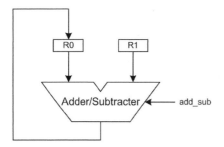

Figure 5.5 A Typical Datapath

Software. The software part of a computer consists of the Operating System, Compiler, and Assembler. The operating system provides an interface between the user and the hardware. The compiler translates the high-level language programs to assembly language programs. The assembler translates the assembly language programs to machine language programs.

ISA. By specifying the format and structure of the instruction set, the ISA specifies the interface between hardware and software of a processing unit. In other words, the ISA provides the details of the instructions that a computer should be able to understand and execute. Each instruction specifies an operation to be performed on a set of data, called *operands*. The operands also show where the result of the instruction should be stored. The *instruction format* describes the specific fields of the instruction assigned to operation and operands. The *opcode* field specifies the operation, and the *operand fields* specify the required data. The way in which the operands can be delivered to an instruction is called *addressing mode*. For example, an operand may be a constant value (*immediate addressing*), contents of a register (*register addressing*), or contents of a memory unit (*direct addressing*). Figure 5.6 shows an instruction format with two operands. It is common to specify some operands explicitly in the instruction and the other operands are implicit. The implicit operands refer to the CPU registers. For example, the instruction *ADD 200*, means "add the content of the memory location *200* with *acc* and store the result in *acc*". Here, *acc* is an *implicit operand* and *200* is an *explicit operand*.

opcode	operand1	operand2

Figure 5.6 Instruction Format

5.3 CPU Design

In the previous two sections, we introduced basic concepts of a computer system. Now we are ready to design a simple CPU, which we refer to as *SMPL-CPU*. Here we describe two different implementations of *SMPL-CPU*, called *Single-Cycle* and *Multi-Cycle* implementation.

5.3.1 CPU Specification

CPU design begins with CPU specification, including the number of *general purpose registers*, *memory organization*, *instruction format*, and *addressing modes*. Note that a CPU is defined according to the application it will be used for.

CPU External Busses. The *SMPL-CPU* has a 16-bit external data bus and a 13-bit address bus. The address bus connects to the memory in order to address locations that are being read or written into. Data read from the memory are instructions and instruction operands, and data written into the memory are instruction results and temporary information. The CPU also communicates with its IO devices through its external busses. The address bus addresses a specific device or device register, while the data bus contains data that is to be written or read from the device.

General Purpose Registers. The *SMPL-CPU* has a 16-bit register, called *accumulator* (*acc*). The *acc* register plays an important role in this CPU. All data transfers, and arithmetic-logical instructions use *acc* as an operand. In a real CPU, there may be multiple accumulators, or a general purpose register-file, each of its registers working like an accumulator.

Memory Organization. The *SMPL-CPU* is capable of addressing 8192 words of memory; each word has a 16 bit width. We assume the memory read and write operations can be done synchronous with the CPU clock in one clock period. Reading from the memory is done by putting the address of the location that is being read on the address bus and issuing the memory read signal. Writing into the memory is done by assigning the right address to the address bus, putting data that is to be written on the data bus, and issuing the memory write signal.

Instruction Format. Each instruction of *SMPL-CPU* is a 16-bit word and occupies a memory word. The instruction format of the *SMPL-CPU*, as shown in Figure 5.7, has an *explicit operand* (the memory location whose address is specified in the instruction), and an *implicit operand* (*acc*). The *SMPL-CPU* has a total of 8 instructions, divided into three classes: *arithmetic-logical instructions* (*ADD*, *SUB*, *AND*, and *NOT*), *data-transfer instructions* (*LDA*, *STA*), and *control-flow instructions* (*JMP*, *JZ*).

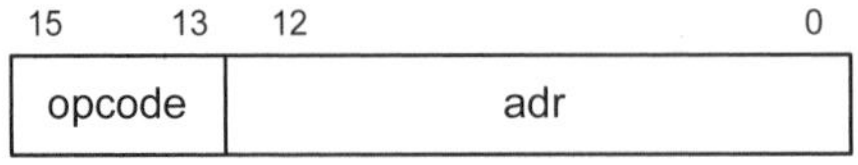

```
15      13   12                                  0
┌────────┬───────────────────────────────────┐
│ opcode │                adr                 │
└────────┴───────────────────────────────────┘
```

Figure 5.7 SMPL CPU Instruction Format

SMPL-CPU instructions are described below. A tabular list and summary of this instruction set is shown in Table 5.1.

- *ADD adr*: adds the content of the memory location addressed by *adr* with *acc* and stores the result in *acc*.

- *SUB adr*: subtracts the content of the memory location addressed by *adr* from *acc* and stores the result in *acc*.

- *AND adr*: ANDs the content of the memory location addressed by *adr* with *acc* and stores the result in *acc*.

- *NOT adr*: negates the content of the memory location addressed by *adr* and stores the result in *acc*.

- *LDA adr*: reads the content of the memory location addressed by *adr* and writes it into *acc*.

- *STA adr*: writes the content of *acc* into the memory location addressed by *adr*.

- *JMP adr*: jump to the memory location addressed by *adr*.

- *JZ adr*: jump to the memory location addressed by *adr* if *acc* equals zero.

Addressing Mode. The *SMPL-CPU* uses direct addressing. For an instruction that refers to the memory, the memory location is its explicit operand and *acc* is its implicit operand.

Table 5.1 SMPL CPU Instruction Set

Opcode	Instruction	Instruction Class	Description
000	ADD *adr*	Arithmetic-Logical	*acc* ← *acc* + Mem[*adr*]
001	SUB *adr*	Arithmetic-Logical	*acc* ← *acc* - Mem[*adr*]
010	AND *adr*	Arithmetic-Logical	*acc* ← *acc* & Mem[*adr*]
011	NOT *adr*	Arithmetic-Logical	*acc* ← NOT (Mem[*adr*])
100	LDA *adr*	Data-Transfer	*acc* ← Mem[*adr*]
101	STA *adr*	Data-Transfer	Mem[*adr*] ← *acc*
110	JMP *adr*	Control-Flow	Unconditional jump to *adr*
111	JZ *adr*	Control-Flow	Conditional jump to *adr*

5.3.2   Single-Cycle Implementation – Datapath Design

Datapath design is an incremental process, at each increment we consider a class of instructions and build up a portion of the datapath which is required for execution of this class.   Then we combine these partial datapaths to generate the complete datapath.   In these steps, we decide on the control signals that control events in the datapath.   In the design of the datapath, we are only concerned with how control signals affect flow of data and function of data units in the data path, and not how control signals are generated.

Step 1: Program Sequencing. The instruction execution begins with reading an instruction from the memory, called *Instruction Fetch* (*IF*).   So an *instruction memory* is needed to store the instructions.   We also need a register to hold the address of the current instruction to be read from the *instruction memory*. This register is called *program counter* or *pc*.   When an instruction execution is completed, the next instruction (which is in the next memory location) should be read and executed.   After the completion of the current instruction, the *pc* should be incremented by one to point to the next instruction in the *instruction memory*.   This leads us to use an adder to increment the *pc*.   Because the size of the memory is 8192 (= 2^{13}) words, the *pc* should be a 13-bit register.   Figure 5.8 shows the portion of the datapath that is used for program sequencing.

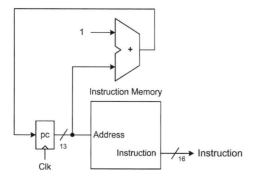

Figure 5.8 Program Sequencing Datapath

Step 2: Arithmetic-Logical Instruction Data-Path. All the arithmetic and logical instructions (except *NOT*) need two operands.   The first operand is *acc*, and the second operand should be read from a memory, called the *data memory*.   The *adr* field of the instruction points to the memory location that contains the second operand.   The result of the operation will be stored in *acc*.   We need a combinational circuit, *arithmetic-logical unit* (*alu*), which performs the operation on the operands of arithmetic and logical instructions.

According to the instruction, the *alu* operation will be controlled by a 2-bit input, *alu_op*.   As shown in Figure 5.9, the *alu* is able to perform addition, subtraction, logical AND, and logical NOT.   The *alu* is designed as other combinational circuits using the methods presented in Chapter 2.

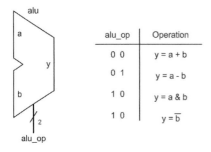

Figure 5.9 SMPL-CPU Arithmetic Logic Unit

Figure 5.10 shows the arithmetic-logical instruction data-path. Note that the address input of the *data memory* comes from *adr* field of the instruction. At this point in our incremental design, the *data memory* needs no control signals, because in all instructions of the type we have considered so far (arithmetic or logical), an operand should be read from the *data memory*. Likewise, this increment of the design does not call for any control signals for writing into *acc*.

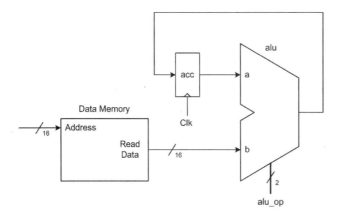

Figure 5.10 Arithmetic-Logical Instructions Data-path

Step 3: Combining the Two Previous Datapaths. Combining the two datapaths constructed so far, leads us to connect the address input of the *data memory* to the *adr* field (bits 12 to 0) of the instruction which is read from the *instruction memory*. This combined datapath is able to sequence the program and execute arithmetic or logical instructions. Figure 5.11 shows the combined datapath.

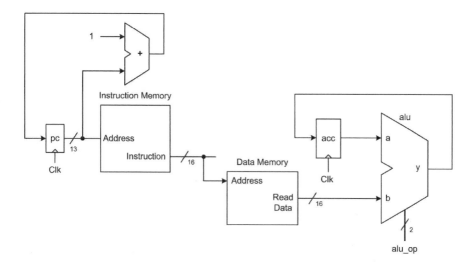

Figure 5.11 Combined Datapath for Program Sequencing & ALU Instructions

Step 4: Data-Transfer Instruction Datapath.   There are two data-transfer instructions in *SMPL-CPU*, *LDA* and *STA*.   The *LDA* instruction uses the *adr* field of the instruction to read a 16-bit data from the *data memory* and store it in the *acc* register.   The *STA* instruction writes the content of *acc* into a *data memory* location that is pointed by the *adr* field.   Figure 5.12 shows the datapath that satisfies requirements of data-transfer instructions.   Because *LDA* reads from the *data memory* while *STA* writes into it, the *data memory* must have two control signals, *mem_read* and *mem_write* for control of reading from it or writing into it.   In data-transfer instructions, only *LDA* writes into the *acc*. When executing an *STA* instruction, *acc* should be left intact.   Having a register without a clock control causes data to be written into it with every clock.   In order to control this clocking, the *acc_write* (write-control, or clock enable) signal is needed for the *acc* register.

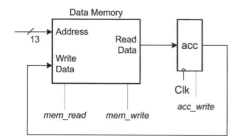

Figure 5.12 Datapath for the Data-Transfer Instructions

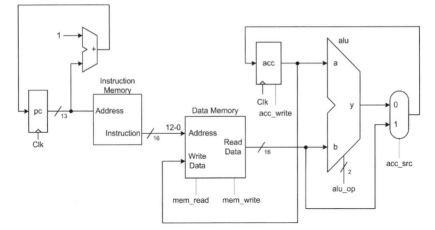

Figure 5.13 Combined Datapath for Program Sequencing, ALU and Data-Transfer

Step 5: Combining the Two Previous Datapaths. Combining the two datapaths, may result in multiple connections to the input of an element. For example, in *Step 3* (Figure 5.10 and Figure 5.11) the *alu* output is connected to the *acc* input and in *Step 4* (Figure 5.12) the *data memory* output (*ReadData*) is connected to the *acc* input. To have both connections, we need a multiplexer (or a bus) to select one of the *acc* sources. When the multiplexer select input, *acc_src* is **0** the *alu* output is selected, and when *acc_src* is **1** the *data memory* output is selected. Figure 5.13 shows the combined datapath.

Step 6: Control-Flow Instruction Datapath. There are two control-flow instructions in *SMPL-CPU*, *JMP*, which is an unconditional jump, and *JZ* which is a conditional jump. The *JMP* instruction writes the *adr* field (bits 12 to 0) of an instruction into *pc*. The *JZ* instruction writes the *adr* field into *pc* if *acc* is zero. So we need a path between the *adr* field of the instruction and the *pc* input. The required datapath is shown in Figure 5.14. As for checking for the zero value of *acc*, a *NOR* gate on the output of this register generates the proper signal that detects this condition. This signal is used for execution of the *JZ* instruction.

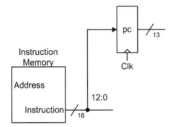

Figure 5.14 Control-Flow Instructions Datapath

Step 7: Combining the Two Previous Datapaths. The increment of *Step 6* created another partial datapath for satisfying operations of our CPU. In this step we are to combine the result of the last step with the combined datapath of Figure 5.13. Considering these two partial datapaths, there are two sources for the *pc* register. One was created in *Step 1*, shown in Figure 5.8 and carried over to Figure 5.13 by *Step 5*, and the other was created in *Step 6* that is shown in Figure 5.14. As in the case of *acc*, we need a multiplexer in the combined datapath to select the appropriate source for the *pc* input. The multiplexer select input is called *pc_src*. If this control signal is **0**, the increment of *pc* is selected and if it is **1**, the address field of the instruction being fetched will be selected. Figure 5.15 shows the combined datapath. This step completes the datapath design of *SMPL-CPU*.

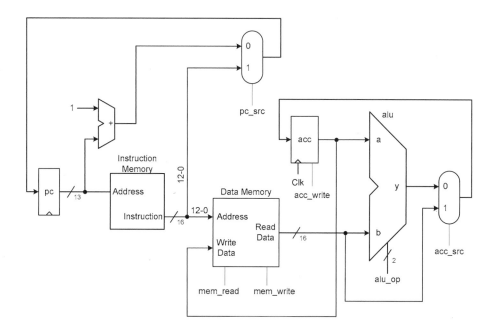

Figure 5.15 The *SMPL-CPU* Datapath

Instruction Execution. Now that we have a complete datapath, it is useful to show how a typical instruction, e.g., *ADD 100*, will be executed in the *SMPL-CPU* datapath. On the rising edge of the clock, a new value will be written into *pc*, and *pc* points the *instruction memory* to read the instruction *ADD 100*. After a short delay, the memory read operation is complete and the controller starts to decode the instruction. Instruction decoding is the process of controller deciding what control signals to activate in order to execute the given instruction. According to this, the controller will issue the appropriate control signals to control the flow of data in the datapath. When data propagation is completed in the datapath, on the next rising edge of the clock, the *alu* output is written into *acc* to complete the execution of the current instruction and the *pc+1* is written into *pc*. This new value of *pc* points to the next instruction. Because the execution of the instruction is completed in one clock cycle, the implementation is called *single-cycle* implementation.

5.3.3 Single-Cycle Implementation – Controller Design

As described before, controller issues the control signals based on the *opcode* field of the instruction. On the other hand, the *opcode* field will not change while the instruction is being executed. Therefore, the control signals will have fixed values during the execution of an instruction, and consequently the controller will be implemented as a combinational circuit.

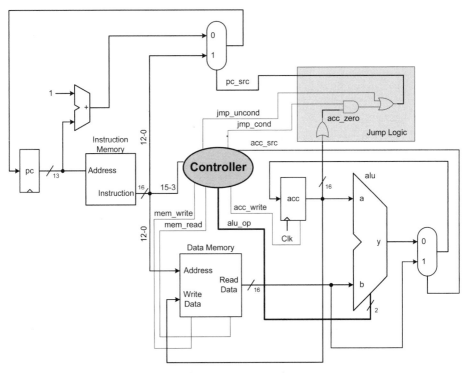

Figure 5.16 Datapath and Controller Interconnection

  Figure 5.16 shows the interconnection of the datapath and controller. As shown in this figure, the controller issues all control signals directly, except *pc_src*, which is issued using a simple logic circuit. For all instructions, except *JMP* and *JZ*, both *jmp_uncond* and *jmp_cond* signals are **0**. With these conditions, the Jump Logic block shown in Figure 5.16 produces a **0** on *pc_src* that causes *pc* to increment. For the *JMP* instruction, the *jmp_uncond* signal becomes **1**, and this puts a **1** on the *pc_src* and directs the *adr* field of the instruction into the *pc* input. For the *JZ* instruction, the *jmp_cond* signal is asserted and if the *acc_zero* signal is **1** (when all bits of *acc* are **0**, the *acc_zero* signal becomes **1**), the address field of the instruction from the *instruction memory* is put into the *pc* register. While executing the *JZ* instruction, if *acc* is not **0**, the *pc+1* source of *pc* is selected.

Table 5.2 Controller Truth Table

Instruction Class	Inst	opcode	mem read	mem write	acc write	acc src	alu op	jmp uncond	jmp cond
Arithmetic Logical	ADD	000	1	0	1	0	00	0	0
	SUB	001	1	0	1	0	01	0	0
	AND	010	1	0	1	0	10	0	0
	NOT	011	1	0	1	0	11	0	0
Data Transfer	LDA	100	1	0	1	1	XX	0	0
	STA	101	0	1	0	X	XX	0	0
Control Flow	JMP	110	0	0	0	X	XX	1	0
	JZ	111	0	0	0	X	XX	0	1

As described before, the controller of a single-cycle implementation of a system is designed as a combinational circuit. In Chapter 2 you learned how to specify a combinational logic using a truth table.   Table 5.2 shows the truth table of the controller.   The values shown are based on activities and flow of data in the datapath.   In what follows, we indicate status of control signals as they are necessary for controlling the flow of data in the datapath.

- **Arithmetic-Logical Class:**
  - o *mem_read=1* to read an operand from the *data memory*.
  - o *acc_write=1* to store the *alu* result in *acc*.
  - o *alu_op* becomes *00, 01, 10,* or *11* depending on the type of arithmetic or logical instruction, i.e., *ADD, SUB, AND,* and *NEG* (see Figure 5.9).
  - o *acc_src=0*, to direct *alu* output to the *acc* input.
  - o *jmp_cond*, and *jmp_uncond* are **0** to direct *pc+1* to the *pc* input.

- **Data-Transfer Class:**
  - o *LDA* Instruction:
    - ▪ *mem_read=1* to read an operand from the *data memory*.
    - ▪ *acc_write=1*, to store the *data memory* output in *acc*.
    - ▪ *alu_op* is *XX*, because *alu* has no role in the execution of *LDA*.
    - ▪ *acc_src=1*, to direct *data memory* output to the *acc* input.
    - ▪ *jmp_cond*, and *jmp_uncond* are **0** to direct *pc+1* to the *pc* input.
  - o *STA* Instruction:
    - ▪ *mem_write=1* to write *acc* into the *data memory*.
    - ▪ *acc_write=0*, so that the value of *acc* remains unchanged.
    - ▪ *alu_op* is *XX* because *alu* has no role in the execution of *STA*.
    - ▪ *acc_src=X*, because *acc* clocking is disabled and its source is not important.
    - ▪ *jmp_cond*, and *jmp_uncond* are **0** to direct *pc+1* to the *pc* input.

- **Control-Flow Class:**

  o *JMP* Instruction:

    - *mem_read* and *mem_write* are **0**, because *JMP* does not read from or write into the *data memory*.
    - *acc_write=0*, because *acc* does not change during *JMP*.
    - *alu_op* is *XX* because *alu* has no role in execution of *JMP*.
    - *acc_src=X*, because *acc* clocking is disabled and its source is not important.
    - *jmp_cond=0*, *jmp_uncond=1*, this puts a **1** on *pc_src* signal and directs the jump address (bits 12 to 0 of instruction) to the *pc* input.

  o *JZ* Instruction:

    - *mem_read* and *mem_write* are **0**, because *JZ* does not read from or write into the *data memory*.
    - *acc_write=0*, because *acc* does not change during *JZ*.
    - *alu_op* equals to *XX*, because *alu* has no role in execution of *JZ*.
    - *acc_src=X*, because *acc* clocking is disabled and its source is not important.
    - *jmp_cond=1*, *jmp_uncond=0*, if *acc_zero* is **1**, this puts a **1** on *pc_src* and directs the jump address (bits 12 to 0 of instruction) to the *pc* input. Otherwise the value of *pc_src* is **0** and *pc+1* is directed to the *pc* input.

5.3.4   Multi-Cycle Implementation

In the single-cycle implementation of *SMPL-CPU*, we used two memory units, two functional units (the *alu* and the adder). To reduce the required hardware, we can share the hardware within the execution steps of an instruction. This leads us to a multi-cycle implementation of *SMPL-CPU*. In a multi-cycle implementation, each instruction will be executed in a series of steps; each takes one clock cycle to execute.

Datapath Design. We start from the single-cycle datapath and try to use a single memory unit which stores both instructions and data, and also a single *alu* which plays the role of both *alu* and the adder. Sharing hardware adds one or more registers to store the output of that unit to be used in the next clock cycle.

To use a single memory, we need a multiplexer to choose between the address of the memory unit from the *pc* output (to address instructions) and bits 12 to 0 of the instruction (to address data). To use a single logic unit instead of the present *alu* and the adder, we should use two multiplexers at the *alu* inputs. The multiplexer on the first *alu* input, chooses between *pc* and *acc*, and the multiplexer at the second *alu* input chooses between memory output and a constant value of 1. Note that *alu* inputs are 16 bits wide, so we append 3 zeros on the left of *pc* to make it a 16-bit vector for the input of the *alu*.

To store the instruction which is read from the memory, a register is used at the output of the memory unit, called *instruction register* (*ir*). The multi-cycle implementation of datapath is shown in Figure 5.17. As mentioned before, the instruction is broken into a series of steps; each takes one clock cycle to execute. These steps are described below:

- **Instruction Fetch (IF)**: In this step we read the instruction from memory and increment *pc*. To do so, we should use *pc* to address the memory, perform a read operation from the memory and write the instruction into *ir*. Also we should apply *pc* to the first *alu* input, the constant value 1 to the second *alu* input, perform an addition, and store the *alu* output in *pc*.

- **Instruction Decode (ID)**: In this step the controller decodes the instruction (which is stored in *ir*) to issue the appropriate control signals.

- **Execution (EX)**: The datapath operation in this step is determined by the instruction class:

  o Arithmetic-Logical Class: Apply bits 12 to 0 of *ir* to the memory, and perform a memory read operation. Apply *acc* to the first *alu* input, and the memory output to the second *alu* input, perform an *alu* operation (addition, subtraction, logical and, and negation), and finally store the *alu* result into *acc*.

  o Data-Transfer Class: Apply bits 12 to 0 of *ir* to the memory. For the *LDA* instruction, perform a memory read operation, and write the data into *acc*. For the *STA* instruction, perform a memory write operation to write *acc* into the memory.

  o Control-Flow Class: For the *JMP* instruction write bits 12 to 0 of *ir* to *pc*. For *JZ*, write bits 12 to 0 of *ir* to *pc* if the content of *acc* is zero.

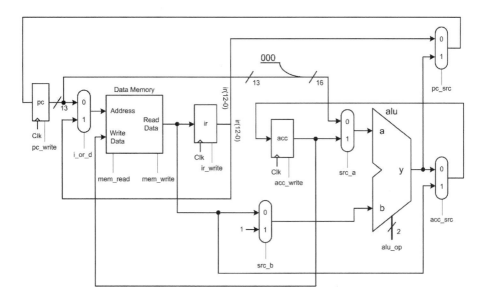

Figure 5.17 SMPL CPU Multi-Cycle Data-Path

Controller Design.   As mentioned before, in multi-cycle implementation of *SMPL-CPU* each instruction is executed in a series of steps. So the controller should specify the appropriate control signals, which have different values for each step. As a result, the controller of a multi-cycle datapath should be designed as a sequential circuit.

Figure 5.18 shows the interconnection of the datapath and controller. As shown in this figure, the controller issues all control signals directly, except *pc_write*, which is issued using a simple logic circuit.

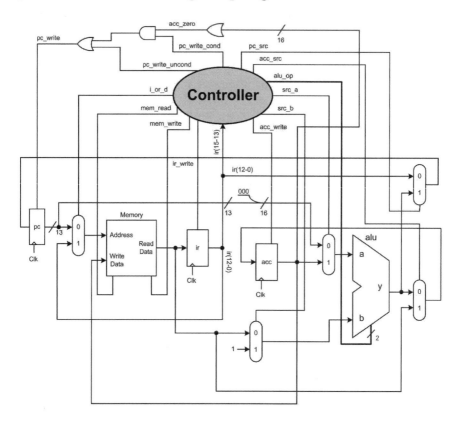

Figure 5.18 Datapath and Controller Interconnection

Figure 5.19 shows the implementation of *SMPL-CPU* controller, which is designed as a Moore finite state machine. As shown in the figure, each state issues appropriate control signals and specifies the next state. There are two important points to be considered, first, transition between states are triggered by the edge of the clock, and second, all control signals in a state are issued by entering the state.

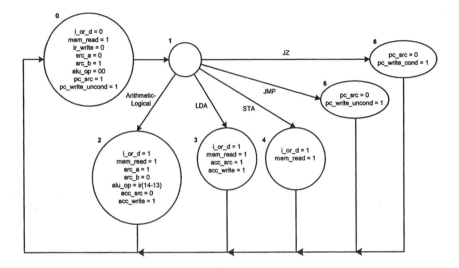

Figure 5.19 SMPL-CPU Multi-Cycle Controller

Instruction execution begins in State 0, which corresponds to the IF step. In this state, *pc* is applied to the memory address input (*i_or_d=0*), the instruction is read from the memory (*mem_read=1*), the instruction is written into *ir* (*ir_write=1*), and *pc* is incremented by 1 (*src_a=0, src_b=1, alu_op=00, pc_src=1, pc_write_uncond=1*). The next state is State 1 that is the step. In this state we give enough time to the controller to decode the instruction, so there is no need to assert any control signal. When the instruction decoding is complete, it specifies the next state according to the type of the instruction being executed.

- Arithmetic-logical instruction: bits 12 to 0 of *ir* are applied to the memory address input (*i_or_d=1*), data is read from the memory (*mem_read=1*), the *acc* output and memory output are directed to the *alu* inputs (*src_a=1, src_b=0*), an *alu* operation is performed (*alu_op=ir[14-13]*), the *alu* output is selected as *acc* input (*acc_src=0*), and the *alu* output is written into *acc* (*acc_write=1*). This finishes the arithmetic-logical instruction execution, and control goes to State 0 to fetch the next instruction.

- LDA instruction: bits 12 to 0 of *ir* are applied to the memory address input (*i_or_d=1*), data is read from the memory (*mem_read=1*), the memory output is selected as the *acc* input (*acc_src=1*), and the memory output is written into *acc* (*acc_write=1*). This finishes the LDA instruction execution, and control goes to State 0 to fetch the next instruction.

- STA instruction: bits 12 to 0 of *ir* are applied to the memory address input (*i_or_d=1*), and *acc* is written into the memory (*mem_write=1*). This finishes the STA instruction execution, and control goes to State 0 to fetch the next instruction.

- *JMP* instruction: bits 12 to 0 of *ir* are written into *pc* (*pc_src=0*, *pc_write_uncond=1*). This finishes the *JMP* instruction execution, and control goes to State 0 to fetch the next instruction.

- *JZ* instruction: bits 12 to 0 of *ir* are written into *pc* if the content of *acc* is zero (*pc_src=0*, *pc_write_cond=1*). This finishes the *JZ* instruction execution, and control goes to State 0 to fetch the next instruction.

The implementation of the controller as discussed above and as shown in Figure 5.19 requires a state machine that can be implemented by a one-hot state machine or an encoded machine with three flip-flops. We will leave this detailed implementation as an exercise for the interested reader.

5.4 Summary

In this chapter we discussed processing units and presented a method of designing general purpose computers. The discussion on the CPU components and its hardware and software was brief and its only purpose was to prepare the reader for the second part of the chapter that discussed the design of a CPU. In presenting the design methodology, we used a simple processor and developed its hardware in several incremental steps. This presentation familiarizes the reader with hardware details of complex CPU architectures and prepares the reader for the CPU example that we will present in the second part of this book.

Part 2

Design Prototyping

This part shows how the Quartus II environment and its related tools are used for entering a design testing it and programming Altera programmable devices. We will show how Altera's UP2 development board can be used for prototyping digital components. Topics covered here are:

- Tools for Design and Prototyping
- Gate Level Combinational Design
- Designing Library Components
- Design Reuse
- HDL Based Design

6 Tools for Design and Prototyping

This chapter introduces tools and environments offered by Altera for design development and prototyping. We will discuss how Altera's Quartus II is utilized to complete a design and program a programmable device. Using UP2 development board for prototyping a design that is developed in Quartus II will be discussed. The chapter also discusses the role of an HDL simulator in design. We will only show the general features of Quartus II and will not get into its details. Various ways Quartus II can be used for design development will be discussed in the chapters that follow.

6.1 Design with Quartus II

This section shows steps involved in using Quartus II for design entry, simulation and device programming. We use a simple AND gate for illustrating the necessary steps.

The environment we are using has several "*Utility Windows*" and "*Tool Bars*". *Utility Windows* are selected by the user and they remain active until they are turned off. On the other hand, *Tool Bars* become active depending on the application you are using. By default, the *Project Manager*, *Status*, and *Messages* windows are on. As shown in Figure 6.1, using the *View* tab and, in there, selecting the *Utility Windows* tab brings up a pull-down menu that enables you to select your active *Utility Windows*.

The *Project Manager* window is a useful window and should be turned on at all times. Project files, design files, their hierarchies, and their compilation status are displayed in this window.

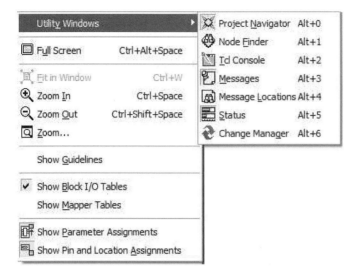

Figure 6.1 Utility Windows

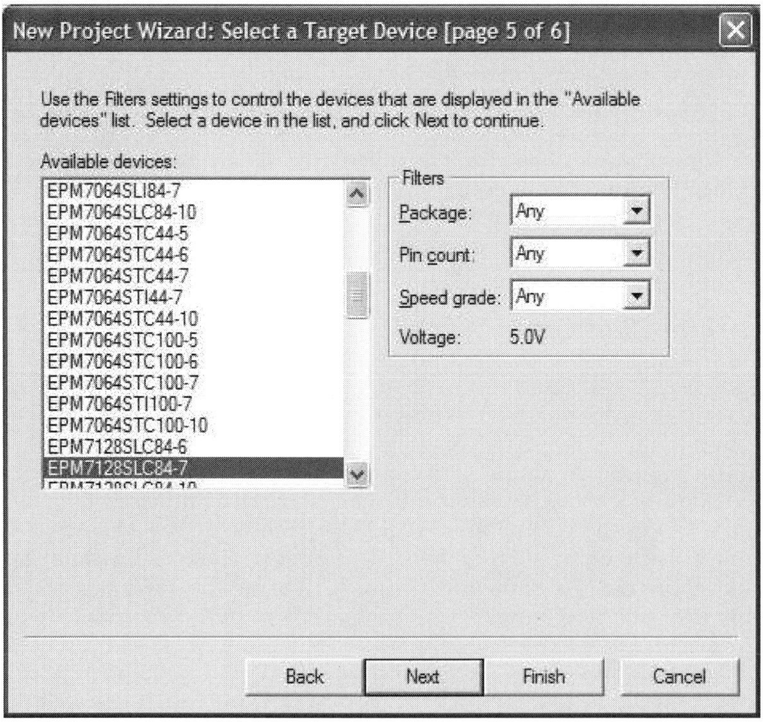

Figure 6.2 Page 5 of Project Definition Pages

6.1.1 Project Definition

The first step in definition of a design in Quartus II is to define a project. A project encloses design files in a given subdirectory. In a project, we define a top-level design that can be simulated and synthesized.

To define a project, use the *New Project Wizard* in the *File* menu. The project wizard brings up a series of windows (pages) for defining project subdirectory, design libraries, tools used with the project, device family, and the specific device used in the project.

For our demo project (the AND gate design), we use the *MyFirstProject* subdirectory, use the *anding* design name, use the MAX7000s device family, and in this family of devices we use the EPM7128SLC84-7 CPLD. We have selected this device because it is one of the two programmable devices on Altera's UP2 development board. Figure 6.2 shows the project definition page in which the project device is defined. After project definition is complete, the name of the project (*anding*) will be displayed in the *Project Navigator* window.

6.1.2 Design Entry

The next step in design development is entering the design. For this purpose, Quartus II offers ways for schematic entry, HDL entry, use of Megafunctions, use of existing parts, tabular specification of functions, and a mixture of all these methods. Entering a design begins by selecting a design entry method from the *Applications* tool bar.

In our demo design we use the schematic entry method, the icon for which appears on the *Applications* tool bar shown in Figure 6.3. Selecting this icon (the one with the little AND-gate), opens a blank schematic file. Next to the schematic file, the *Block and Symbol Editors* tool bar opens that allows selection of gates and components for our new schematic file.

Figure 6.3 *Applications* **Tool Bar**

Figure 6.4 *Block and Symbol Editors* **Tool Bar**

To enter components of your design, double-click anywhere on the schematic window, or select the *Symbol Tool* (the little AND-gate) from the *Block and Symbol Editors* tool bar of Figure 6.4. This opens the *Symbol* window in which available libraries, including the standard Quartus II library, are shown.

Open this library by clicking on the little plus sign next to it. Once opened, select *Primitives*, and in the *Logic* hierarchy that opens select *and2*. This selection is shown in the *Symbol* window of Figure 6.5. Alternatively, if you already know the name of the component you are using, enter it in the *Name* area shown in this figure.

The window of Figure 6.5 closes by clicking *OK*. When this closes, the selected symbol becomes available for placement on your schematic window. Click anywhere on this window to place your AND gate. The schematic window of Figure 6.6 shows this AND gate.

When a design is complete, you have to define its input and output pins. Since in our case, our design is just a simple AND gate, we now have to enter its pins and connect them to core of our design (the AND gate). For this purpose, enter pins the same way you entered your AND gate.

In the library hierarchy of the *Symbol* window of Figure 6.5, open the *pins* folder, select *input* and *output* and place two of the former and one of the latter in your schematic window of Figure 6.6. This figure shows all design components that must now be connected.

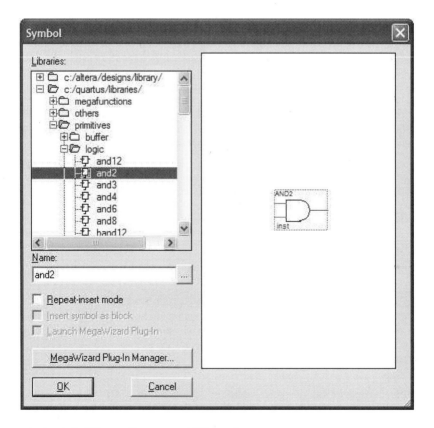

Figure 6.5 Library Component Hierarchy

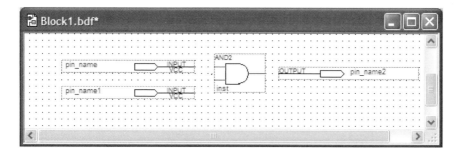

Figure 6.6 Schematic Window

Figure 6.7 Completed *anding* Block Diagram File

To connect components of Figure 6.6, in the *Block and Symbol Editors* tool bar of Figure 6.4, select the *Orthogonal Node Tool* (the 90° thin line with dots on its ends). This makes your curser a wiring tool that can be used to connect nodes together. Use this tool to connect the input pins to the inputs of the AND gate and the output pin to the gate's output. When done, disable the wiring tool by clicking the arrow on the tool bar of Figure 6.4.

Before completing your design, rename input and output ports to names you are comfortable with. We use *a*, *b*, and *w* for the inputs and the output of our design. To name a pin, either double-click it to open its *Pin Properties* window, or right-click it and select *Properties* from the pull-down menu that shows up.

When this is completed, save your design and make sure it is named the same as your project, i.e., *anding*. Figure 6.7 shows the completed block diagram of our *anding* design. The top level entity of a design must be named the same as its project.

6.1.3 Device Configuration

In the next step of our design development, we use the *Assign Pins* window to assign input and output ports of our design to pins of the CPLD we are using. To bring up this window go to the *Assignments* tab and in the pull-down menu that shows up select *Assign Pins...* . The corresponding window is shown in Figure 6.8.

A straight forward way of assigning pins is to scroll down the device's pin list, and click on the pin that is being assigned to a design node. Then in the *Assignment* area, next to the *Pin name* type the name of the node of your design. Alternatively, you can click on the three dots next to the *Pin name* box to have the *Node Finder* utility find your design's nodes. This feature can only be used if a design has already been compiled. In the *Node Finder* window, clicking *Start* begins the search for nodes of your design. We can assign nodes from our design to the numbered device pin by selecting it from the list of nodes found.

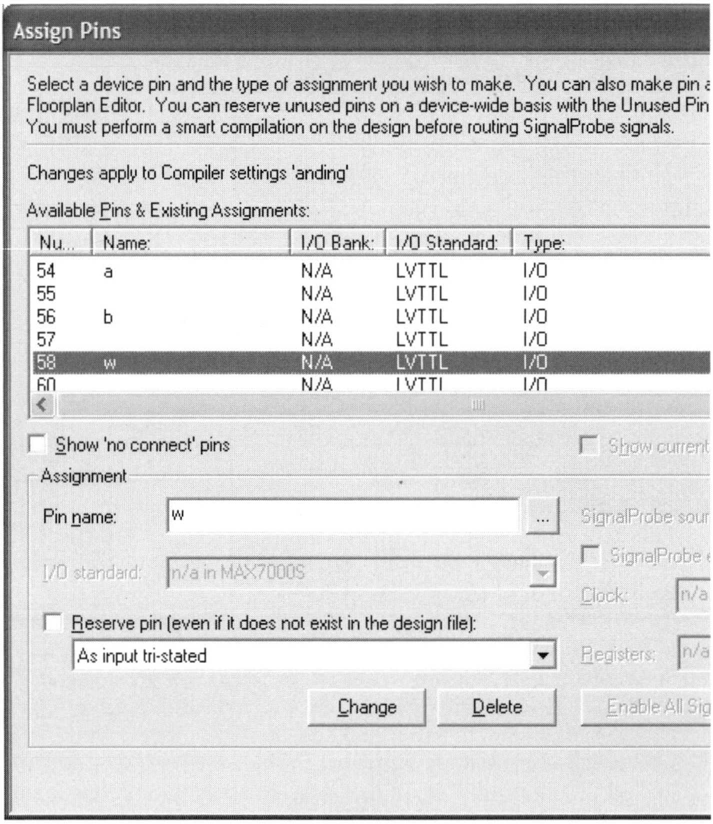

Figure 6.8 *Assign Pins* Window to Assign Deign ports to CPLD Pins (Partial View)

In our design we use the straight forward method of typing the name of the node in association of the highlighted pin number. This is partly due to the fact that we have not compiled our design yet, and node names are not available. For our design, we assign *a*, *b*, and *w* nodes to device pins 54, 56 and 58 respectively. The reason for this selection becomes clear after we discuss the UP2 board in the next section.

6.1.4 Design Compilation

After the completion of the design entry phase, and for simulation and device programming, we have to compile our design. Compilation consists of many phases including, design analysis, synthesis, binding, layout and placement, timing extraction, and building data files for chip layout and floor plan. For our AND gate design many of these phases are either very simple or completely eliminated. For example the synthesis phase that converts our behavioral designs to a netlist of gates, does not have a strong role in compiling our design.

To start the compilation process, select the *Start Compilation* icon from the *Standard* tool bar. This icon is the right-pointed triangle shown in Figure 6.9. When compilation begins, the status of its various phases is displayed in the *Status* utility window. If errors occur, they will be displayed in the *Messages* utility window.

Figure 6.9 *Standard* **Tool Bar**

6.1.5 RTL View

As a result of compilation, Quartus II generates a schematic diagram of the hardware that is to be programmed into a programmable device. This schematic is independent of the target device and uses basic logic gates and primitive functions. To see the RTL view of a compiled design, go to the *Tools* pull-down menu and select *RTL Viewer*.

Figure 6.10 shows the RTL view of our *anding* project. Since the design we are dealing with is a simple AND, the RTL view of it is just an AND gate shown in this figure. However, for a large design, the RTL view shows the design hierarchy and post-synthesis netlist of the components of the design.

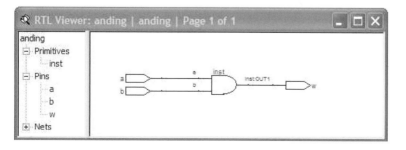

Figure 6.10 RTL View of *anding*

6.1.6   Post-Synthesis Simulation

A design that is compiled can be simulated. This simulation is often referred to as post-synthesis, because it simulates the actual gates and cells of the target device (the device that will be programmed). In our case, no synthesis was necessary because of our simple design. However, we still use this terminology to reflect the detailed simulation that is being performed.

Before we start our simulation, we have to specify input values for the inputs of our design. To do this, select the *New Vector Waveform File* icon (the icon with the square wave) from the *Applications* tool bar of Figure 6.3 to open a new waveform definition file. The waveform file is initially blank and we need to enter our node names and their associated waveforms. Figure 6.11 shows our waveform file after node names are entered and waveforms defined for them.

To enter design nodes, double-click in the area under the word "Name" in the waveform file. Doing this, or right-clicking and selecting the "*Inset Node or Bus...*" brings up the *Insret Node or Bus* window shown in Figure 6.12. In this window, in the white box next to *Name*, type the name of your node, or select the *Node Finder* to find your design's nodes.

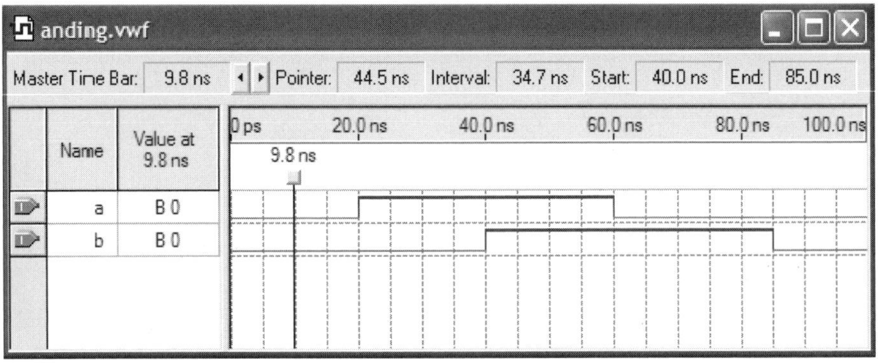

Figure 6.11 Waveform Definition File

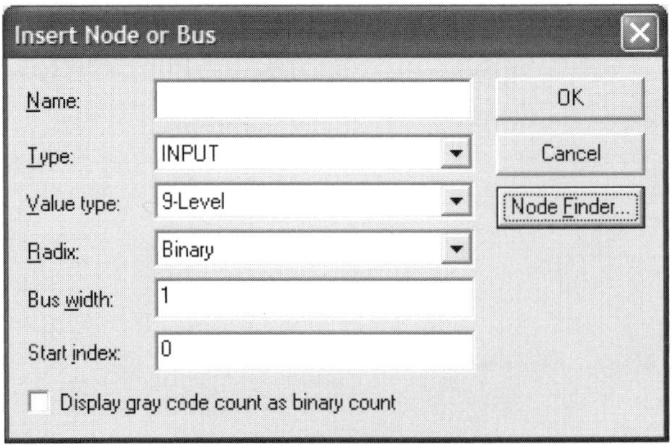

Figure 6.12 *Insert Node or Bus* Window

Figure 6.13 *Waveform Editor* **Tool Bar**

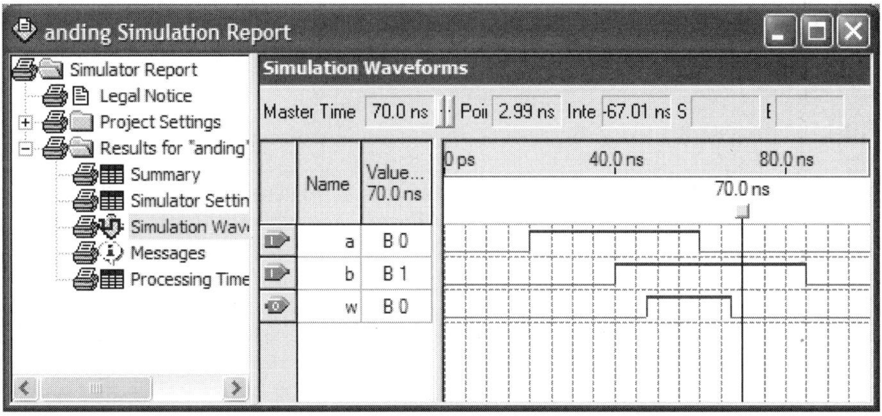

Figure 6.14 Simulation Run Results

In order to define waveforms for our input ports, we can take advantage of the tools provided by the *Waveform Editor* tool bar, shown in Figure 6.13. One way of doing this, is to select the *Waveform Editing Tool* (the icon with crossing waveforms and a two-headed arrow inside it). When this is selected, you can use your mouse to paint your waveform in the waveform area of Figure 6.11 to paint **0**s and **1**s.

Alternatively, you can select the arrow in the *Waveform Editor*, and in your waveform area paint a portion of the waveform. The painted area of the waveform will highlight. Then select one of the icons in the right part of the *Waveform Editor* tool bar (Figure 6.13) to set a value in the highlighted area.

When definition of the waveform is completed, save it as *anding.vwf*. Using the same name as your design, associates this waveform with the *anding* schematic file. This readies everything that we need for our simulation run. To start the simulation, select the simulation icon from the *Standard* tool bar of Figure 6.9 (the simulation icon is the right-pointed triangle with a little waveform underneath it).

While simulation is running it status appears in the *Status* utility window. It will show all processes as 100% when the run is complete. At this time a new waveform file will be displayed that shows the input waveforms as we entered and the resulting output pin. Figure 6.14 shows this simulation report.

Note in this waveform that there is a delay from the time that the *b* input becomes **1** to the time that *w* becomes **1**. This delay (7.761 ns) is due to the delays of logic cells of our CPLD and the delay of its IO cells. A successful simulation run verifies the correctness of our design as well as the timing of the input waveforms. If we change our input waveforms too fast, the outputs cannot catch up and we will end up with hard-to-justify output waveforms.

6.1.7   Device Programming

The last step in developing a design for a PLD is programming the target PLD. The necessary files for this purpose are generated in Quartus II after a design has been successfully compiled. MAX 7000S devices use the *pof* (Programmer Object File) programming file format and the FLEX 10K devices use *sof* (SRAM Object File). These files that are generated by the compiler, include all necessary configuration data for the appropriate PLDs. The *pof* files configure EEPROMs and *sof* files are loaded in FPGA SRAMs.

To start the device programming process, select the *Programmer* icon from the *Applications* tool bar shown in Figure 6.3 (the icon with some wave on top of a little chip). This brings up the device configuration window, shown in Figure 6.15, and its associated *Programmer* tool bar, shown Figure 6.16. In the configuration window, you specify the hardware used for programming your device, the programming file (a *pof* file for our CPLD, generated by the compiler) and the specific device being programmed.

If your Quartus II environment is being used for the first time, you should setup its programmer hardware. To do this, click on *Hardware* in Figure 6.15 and in the *Hardware Setup* window that opens add your hardware, select it and close the window. In our example of Figure 6.9 we have specified ByteBlasterII that is connected to the LPT1 port of our computer. ByteBlasterII is a device programmer by Altera for Altera devices, and connects to the parallel printer port of your PC.   The UP2 development board uses this hardware for programming its devices. Referring back to Figure 6.15, the selected *Mode* of programming is *JTAG*, which is the way ByteBlasterII connects to UP2.

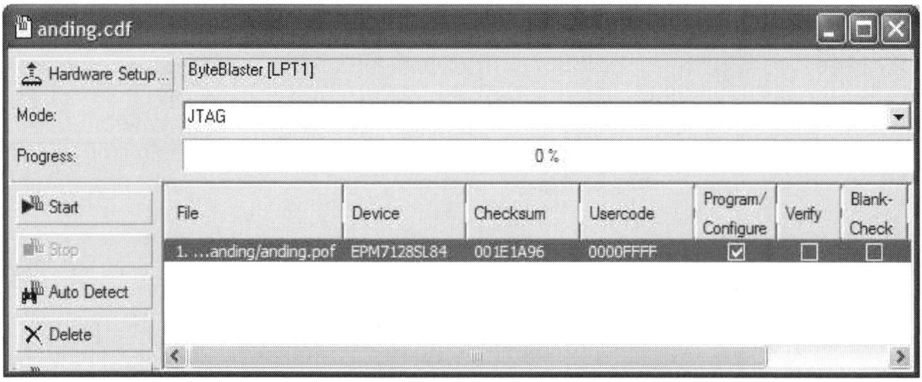

Figure 6.15 Device Programming Window

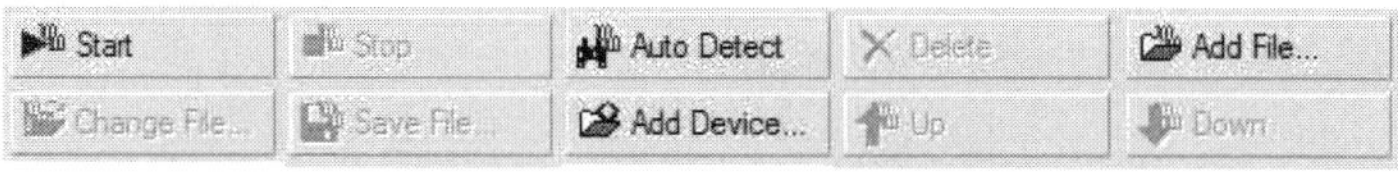

Figure 6.16 *Programmer* **Tool Bar**

If the programming file does not automatically appear in the configuration window of Figure 6.15, or if you need to change the file, double-click in the white space in the *File* area and select the appropriate programming file. Likewise, if the device you are programming is not already shown under the *Device* heading in Figure 6.15, you can select your device by double-clicking this area.

After you have completed your setup in the configuration window, put a check-mark in the *Program/Configure* box and select the *Start Programming* icon from the *Programmer* tool bar of Figure 6.16. The right-pointed triangle with some wave on its right hand side is the icon for starting the device programming process. When programming is being done, the corresponding messages appear in the Messages window and its progress is shown in the configuration window.

6.1.8 Configured Devices

Quartus II enables you to inspect the timing of your configured devices, see their floor plan, and in some cases change the way device cells are used. Three icons in the *Standard* tool bar of Figure 6.9 activate *Timing Closure Floorplan*, *Last Compilation Floorplan*, and *Chip Editor* applications. The timing of ports to the CPLD cells and internal CPLD delays can be seen in the *Timing Closure* window.

Figure 6.17 shows the *Timing Closure* window resulted from compilation of our AND gate design targeting the EPM7128SLC84-7 CPLD. As shown, there are eight PLDs in this CPLD (see Section 4.3). These PLDs are numbered A to H from the upper left to lower right CPLD.

For navigating in this floorplan, there is an associated tool bar that is shown in Figure 6.18. This tool bar becomes active when a floorplan of the device is displayed.

Zooming on the floorplan of Figure 6.17 enables us to see the actual IO pins used for our design, macrocells that have been utilized, and their timings. Figure 6.19 shows PLD "F" that is used for implementing our *anding* design. The arrow from *b* to *w* shows the internal timing delay value.

Other features on the *Timing Closure* window include display of critical times, connection counts, delays in and out of cells, node equations, user specified pin assignments, fitter pin assignments, and other parameters that show how the compiler and the device programmer have implemented our design in our target device. This information is displayed by selecting appropriate icons from the *Floorplan Editor* tool bar of Figure 6.18.

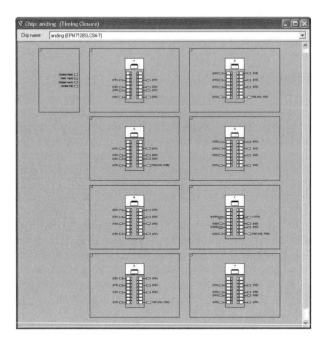

Figure 6.17 Floorplan Showing MAX 7000S Eight PLDs

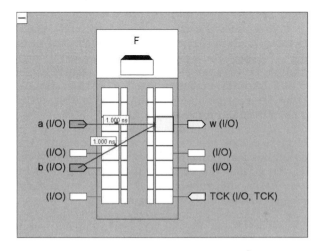

Figure 6.18 *Floorplan Editor* Tool Bar

Figure 6.19 PLD "F" Utilized in MAX 7000S

6.2 Hardware Description Language Based Design

An alternative to using schematic capture for a complete design or components of a design is to use an HDL. The Quartus II software allows describing a component used in the schematic window in VHDL or Verilog. This software also allows a complete design to be described in an HDL. In either case, Quartus II will synthesize the HDL portion of the design, incorporate it with the rest of the design, and generate device layout and appropriate device programming files the same way it does for any other design. This means that Quartus II allows post-synthesis simulation of HDL designs, the same way it allows detailed simulation of design parts using other design entry methods.

The problem with treating an HDL based design the same as other design parts is that, an HDL design is generally more complex and consists of custom new code developed by a designer. That is unlike a purely schematic based design that uses pre-tested existing parts in a hierarchical fashion. Therefore, before a designer uses his or her HDL code with other parts of a design and synthesizes it along with other components, he or she must simulate the HDL design to verify its correct operation.

This means that in addition to all tools and utilities that Quartus II offers, a complete digital design environment needs a high-level HDL simulator. The simulator that we use for this purpose is the ModelSim Altera 5.7c simulator from Mentor Graphics Inc. This simulator that is customized for Altera is distributed with the full version of Quartus II software. Since this simulator is not an integrated part of the Quartus II software, other HDL simulators can also be used for simulating our high-level HDL designs.

In this section we show how a Verilog description written for an "or" function is simulated and tested in ModelSim, and how it is incorporated in Quartus II for synthesis and device programming. Figure 6.20 and Figure 6.21 show the "or" description and its testbench for simulation by ModelSim.

```
`timescale 1 ns / 1 ns
module ored2 (a, b, w);
input a, b;
output w;
  assign w = a | b;
endmodule
```

Figure 6.20 OR Gate Verilog File (Filename is *ored2.v*)

```
`timescale 1 ns / 1 ns
module or2tester();
reg aa, bb;
wire ww;
  ored2 U1 (aa, bb, ww);
  initial begin aa = 0; bb = 0; end
  initial repeat (4) #17 aa = ~ aa;
  initial repeat (3) #29 bb = ~ bb;
endmodule
```

Figure 6.21 OR Gate Testbench (Filename is *or2tester.v*)

When ModelSim begins, it brings up the last design project. The first thing to do for specifying a new design for simulation is creating a new project. For doing this, go to the *File* menu of the main ModelSim window and select *New Project*, as shown in Figure 6.22. This brings up a *Create Project* window (also shown in Figure 6.22), in which the name of the project, its location and its library name can be specified.

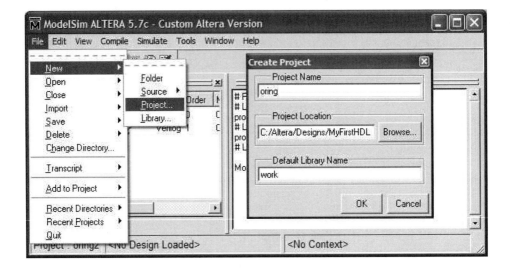

Figure 6.22 Creating a new Project in ModelSim

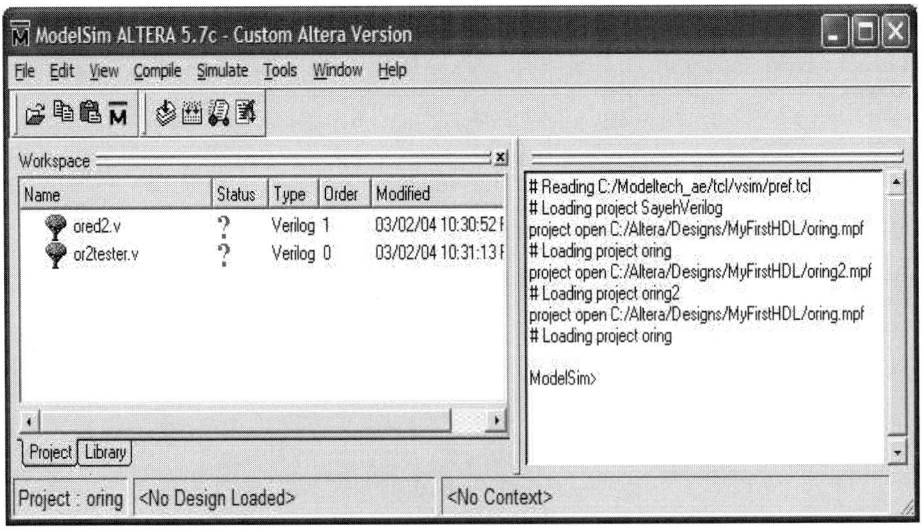

Figure 6.23 Design Files Added to the Project

Following the project creation step, design files must be added to the project. To enter your new design files, start in the *File* menu and follow *File* → *New* → *Source* → *Verilog*. This opens a Verilog editing window. When editing a new file is completed, it is automatically added to your active project.

Alternatively, you can add existing files to your design project by selecting *File* → *Add to Project* → *Existing File ...*, and then selecting an existing file. In our example, we add *ored2.v* and *or2tester.v* files to our *oring* project. Figure 6.23 shows the resulting ModelSim window.

The next step is to compile all our design files. To do this, click on the *Compile* tab and follow *Compile* → *Compile All*. After a successful compilation, the question-marks next to the file names under the status of your design files (Figure 6.23) will be changed to check-marks.

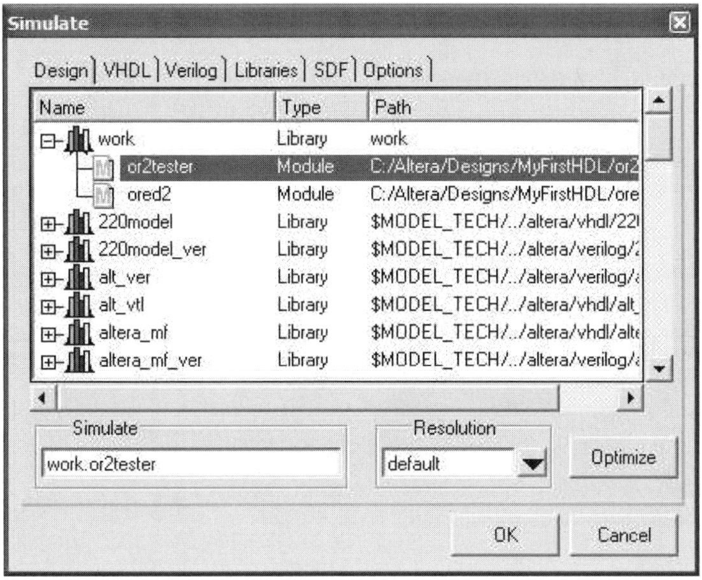

Figure 6.24 Starting the Simulation

The next step after a successful compilation is to initiate the simulation. For this purpose, start from the *Simulate* menu and follow *Simulate* → *Simulate* .... This brings up the *Simulate* window shown in Figure 6.24. Since we have compiled our design files in the *work* library (see *Create Project* window in Figure 6.22), our compiled simulation model is available in this library. Open the *work* library of Figure 6.24, and select the top-level entity of the design that is being simulated. In our example, the *or2tester* module that encloses all other modules is the top-level design entity. Note that you only have to do this once, and successive simulations can just be started by clicking on various simulation icons that will appear in the ModelSim main window.

When the *Simulate* window is Okayed, the simulation initializes. We can now select signals from our design hierarchies to display in the waveform window. To do this, in the *View* pull-down menu select *Signals* and *Wave* windows. This opens the *Signals* and *Wave* windows.

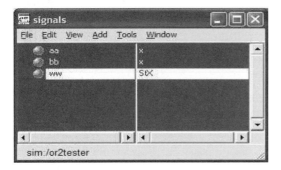

Figure 6.25 *Signals* Window to Select Display Signals

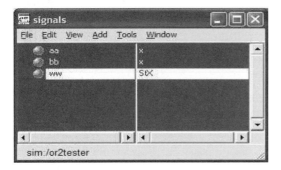

Figure 6.26 *Wave* Window Displaying Simulation Results

The *Signals* window, shown in Figure 6.25, contains signal names of the top-level entity of our design. On the other hand, the *Wave* window is initially blank. Select and copy signals from the *Signals* window and paste them in the *Wave* window. Use the *Edit* tab in both windows for select, copy and paste.

To run the simulation, start from the *Simulate* tab, and in the pull-down menu that appears follow: *Simulate* → *Run* → *Run -All*. This causes the simulation to continue until no more events occur in the design. Our simulation run stops at 87 ns. Figure 6.26 shows our simulation results. If further simulation runs of this testbench become necessary, follow: *Simulate* → *Run* → *Restart...*, and in the *Restart* window that opens select the windows you want to keep (like *Signals* of Figure 6.25 and *Wave* of Figure 6.26), and then click on *Restart*.

6.2.1 Porting to Quartus II

The design that is completed and tested in ModelSim can be used as a component of another design, or a complete design in Quartus II. We now show steps necessary for porting our tested *ored2.v* Verilog description to a Quartus II project, e.g., *oring*.

Copy the tested Verilog file of *ored2* to the directory of the *oring* project. Then, in the *File* menu of Quartus II select *Open* and in the window that opens select the *ored2.v* file. This opens a text editor window containing the *ored2.v* file.

To use this file in another design, you have to generate a symbol for it. The easiest way to generate a symbol for a design file (schematic or HDL design files) is to have Quartus II generate a default symbol. For doing this, while your design file is open, go to the *File* pull-down menu of Quartus II and follow: *File → Create/Update → Create Symbol Files for Current File.* This will generate a symbol for the design file that can be used in the design of our *oring* project.

To use this symbol, in the *Symbol* window that opens when a component is being placed in a schematic file (Figure 6.5), select the *Project* folder and in this folder you will find all symbols that have been added to your current project. Figure 6.27 shows the *Symbol* window of the *oring* project that contains our own *ored2* symbol. This symbol can be selected and used in a schematic file just like any other symbol, e.g., the *AND2* primitive of Section 6.1.2. Figure 6.28 shows the schematic of the *oring* project that uses two of the instances of our own *ored2* symbol and one instance of Quartus II *NAND2* primitive.

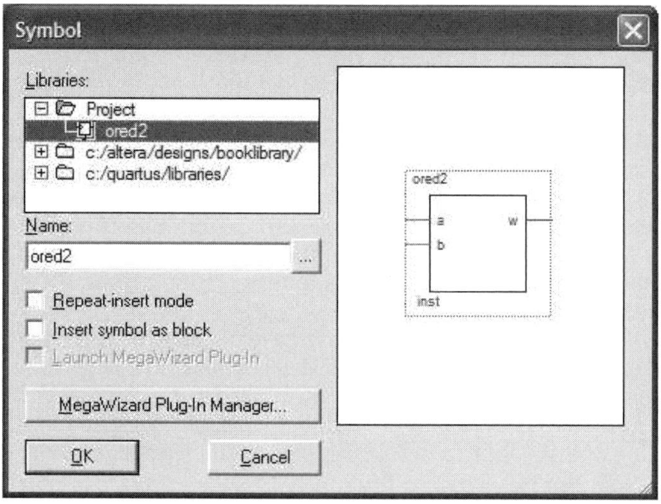

Figure 6.27 Selecting a Symbol from the Current Project Directory

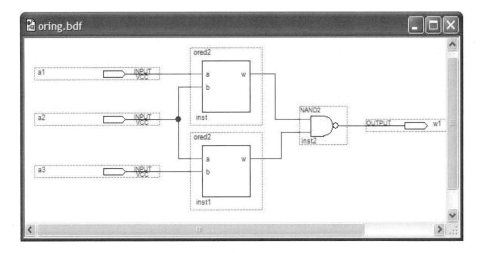

Figure 6.28 Using *ored2* Symbol in *oring* Project

6.3   UP2 Development Board

As a simulation model is a soft model of a hardware that is being designed, a prototype is a hard model of the hardware under design. A development board includes one or more devices to be used as prototypes, and provides easy programming and interfaces for utilization of such devices.   Such a board provides basic I/O devices and interfaces for the peripherals that the prototype devices may be used with.

Having a development board available, a digital designer develops his or her design using simulators at various levels of abstraction, synthesis tools, placement programs, and other related tools.   When the design is complete, devices on a development board are programmed, and using the interfaces provided on the board, the actual peripherals are connected to the prototype device. This provides an exact hardware model of the design that can be tested for actual physical conditions.

Because of the growth of programmable devices, development boards have become very popular, and are available for various devices, various degrees of complexity, and many common interfaces. It is not unusual for a company with a product that is being distributed to a limited number of customers, to use a development board in the product that is being shipped. This way, the system will be tested and examined on-site, and the final board will be generated after all design and board layout bugs are fixed.

Altera's UP2 development board is mainly designed for educational purposes. In a university setting, this board is used for laboratories related to courses on digital system design, computer architecture, and peripheral design. In spite of its main educational target, UP2, with its popular MAX 7000S CPLD and FLEX 10K FPGA, is a very useful development board for industrial settings.

Altera's "*University Program UP2 Development Kit*" datasheet is a document that describes details of this board. In this section of this book, we will discuss some of its feature that we use in the design projects of the rest of this book.

6.3.1   UP2 General Features

The UP2 development board, shown in Figure 6.29, has an EPM7128SLC84-7 CPLD and an EPF10K70RC240-4 FPGA.   The board has an interface for ByteBlaster II device programming hardware that can be used to program the on-board devices. The board provides power and clock for its CPLD and FPGA.

CPLD. The EPM7128S device, a member of the MAX 7000S family, is based on erasable programmable read-only memory (EEPROM) elements. The EPM7128S device features a socket-mounted 84-pin plastic j-lead chip carrier (PLCC) package and has 128 macrocells. With a capacity of 2,500 gates and a simple architecture, the EPM7128S device is ideal for introductory designs as well as larger combinatorial and sequential logic functions.

FPGA. The EPF10K70 device is based on SRAM technology. It is available in a 240-pin RQFP package and has 3,744 logic elements (LEs) and nine embedded array blocks (EABs). With 70,000 typical gates, the EPF10K70 device is ideal for intermediate to advanced digital design courses, including computer architecture, communications, and DSP applications.

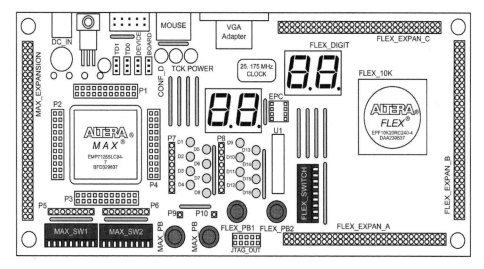

Figure 6.29 UP2 Development Board

ByteBlaster II. Designs can be easily and quickly downloaded into the UP2 board using the ByteBlaster II download cable, which is a hardware interface to a standard parallel port. This cable sends programming or configuration data between the device programming software (programmer part of Quartus II) and the UP2 Education Boards. Because design changes are downloaded directly to the devices on the board, prototyping is easy and multiple design iterations can be accomplished in quick succession.

JTAG Input Header. The 10-pin female plug on the ByteBlaster II download cable connects with the JTAG_IN 10-pin male header on the UP Education Board. The board provides power and ground to the ByteBlaster II download cable. Data is shifted into the devices via the TDI pin and shifted out of the devices via the TDO pin. In all of our configurations we use the ByteBlaster II in the Joint Test Action Group (JTAG) operating mode.

Jumpers. The UP Education Board has four three-pin jumpers (TDI, TDO, DEVICE, and BOARD) that set the JTAG configuration. The JTAG chain can be set for a variety of configurations (i.e., to program only the EPM7128S device, to configure only the FLEX 10K device, to configure and program both devices, or to connect multiple UP Education Boards together). In all of our work with UP2 we only use one board and only one of the programming devices.

Supply Power. The DC_IN power input accepts a 2.5-mm × 5.55-mm female connector. The acceptable DC input is 7 to 9 V at a minimum of 350 mA. The RAW power input consists of two holes for connecting an unregulated power source. After being regulated by the board hardware proper DC voltage is applied to both devices on the board.

Oscillator. The UP Education Board contains a 25.175-MHz crystal oscillator. The output of the oscillator drives a global clock input on the EPM7128S device (pin 83) and a global clock input on the FLEX 10K device (pin 91).

6.3.2   EPM7128S CPLD Device

Resources for the UP2 CPLD device include prototyping headers, switches, push-buttons, LEDs and seven-segment displays. Female connectors surrounding this device give access to its pins for connecting to on-board or external resources.

Prototyping Headers. The EPM7128S prototyping headers are female headers that surround the device and provide access to the device's signal pins. The 21 pins on each side of the 84-pin PLCC package connect to one of the 22-pin, dual-row 0.1-inch female headers. The pin numbers for the EPM7128S device are printed on the UP2 Education Board (an "X" indicates an unassigned pin). Table 6.1 lists the pin numbers for the four female headers: P1, P2, P3, and P4. The power, ground, and JTAG signal pins are not accessible through these headers.

Push-Buttons and Switches. MAX_PB1 and MAX_PB2 are two push-buttons that provide active-low signals and are pulled-up through 10-KΩ resistors. Pins from the EPM7128S device are not pre-assigned to these push-buttons. Connections to these signals are made by inserting one end of the hook-up wire into the push-button female header. The other end of the hook-up wire should be inserted into the appropriate female header assigned to the I/O pin of the EPM7128S device.

Table 6.1 Pin Numbers for Each Prototyping Header

P1		P2		P3		P4	
Outside	Inside	Outside	Inside	Outside	Inside	Outside	Inside
75	76	12	13	33	34	54	55
77	78	14	15	35	36	56	57
79	80	16	17	37	38	58	59
81	82	18	19	39	40	60	61
83	84	20	21	41	42	62	63
1	2	22	23	43	44	64	65
3	4	24	25	45	46	66	67
5	6	26	27	47	48	68	69
7	8	28	29	49	50	70	71
9	10	30	31	51	52	72	73
11	X	32	X	53	X	74	X

MAX_SW1 and MAX_SW2 each contain eight switches that provide logic-level signals. These switches are pulled-up through 10-KΩ resistors. Pins from the EPM7128S device are not pre-assigned to these switches. Connections to these signals are made by inserting one end of the hook-up wire into the female header aligned with the appropriate switch. Insert the other end of the hook-up wire into the appropriate female header assigned to the I/O pin of the EPM7128S device. The switch output is set to logic 1 when the switch is open and set to logic 0 when the switch is closed.

LEDs and Displays. The UP Education Board contains 16 LEDs that can be used with the EPM7128S device. These LEDs are pulled-up with a 330 Ω resistor. An LED is illuminated when a logic 0 is applied to the female header associated with the LED. Pins from the EPM7128S device are not pre-assigned to LEDs. Connections to LEDs are made by inserting one end of the hook-up wire into the LED female header. The other end of the hook-up wire should be inserted into the appropriate female header assigned to the I/O pin of the device. Figure 6.30 shows female headers corresponding to the UP2 LEDs.

MAX_DIGIT is a dual-digit, seven-segment display connected directly to the EPM7128S device. Each LED segment of the display can be illuminated by driving the connected EPM7128S device I/O pin with a logic 0. Figure 6.31 shows the display segments and their connections to EPM7128S pins.

Figure 6.30 LED Corresponding Female Headers

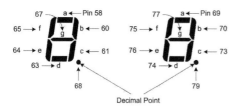

Figure 6.31 EPM7128S Pin Connections to MAX Digits

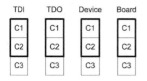

Figure 6.32 Jumper Settings for Programming Only the EPM7128S Device

Programming. EPM7128S of the UP2 is programmed by the ByteBlaster II hardware connected to the JTAG terminal of the board. Board jumpers must be set as shown in Figure 6.32 in order to program this device.

6.3.3 EPF10K70 FPGA Device

Resources for the UP2 FPGA device include switches, push-buttons, seven-segment displays, a VGA connector, and a keyboard/mouse interface connector. Pins from the EPF10K70 device are pre-assigned to these resources. For connection to other peripherals, expansion pins on the sides of the UP2 board should be used.

Push Buttons and Switches. FLEX_PB1 and FLEX_PB2 are two push buttons that provide active-low signals to two general-purpose I/O pins on the FLEX 10K device. FLEX_PB1 connects to pin 28, and FLEX_PB2 connects to pin 29. Each push button is pulled-up through a 10-KΩ resistor. Figure 6.33 shows FLEX pin connections to the two available push-buttons.

FLEX_SW1 contains eight switches that provide logic-level signals to eight general-purpose I/O pins on the FLEX 10K device. An input pin is set to logic 1 when the switch is open and set to logic 0 when the switch is closed. Figure 6.34 shows FLEX pin connections to the FLEX switch set.

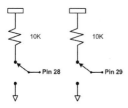

Figure 6.33 FLEX Pin Connections to its Push Buttons

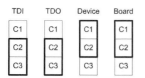

Figure 6.34 FLEX Pin Connections to its Switches

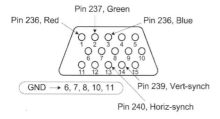

Figure 6.35 FLEX Pin Connections to its Seven-Segment Displays

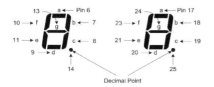

Figure 6.36 FLEX Pin Connections to the VGA D-sub Connector

Pin 30, MOUSE_CLK → 1
Pin 31, MOUSE_DATA → 3

Figure 6.37 FLEX Pin Connections to the Mouse Mini-DIN Connector

TDI	TDO	Device	Board
C1	C1	C1	C1
C2	C2	C2	C2
C3	C3	C3	C3

Figure 6.38 Jumper Settings for Configuring Only the EPF10K70 Device

Displays. FLEX_DIGIT is a dual-digit, seven-segment display connected directly to the FLEX 10K device. Each LED segment on the display can be illuminated by driving the connected FLEX 10K device I/O pin with a logic 0. See Figure 6.35 for FLEX pin connections to the display segments.

VGA Interface. The VGA interface allows the FLEX 10K device to control an external video monitor. This interface is composed of a simple diode-resistor network and a 15-pin D-sub connector (labeled VGA), where the monitor can plug into the boards. The diode-resistor network and D-sub connector are designed to generate voltages that conform to the VGA standard. Information about the color, row, and column indexing of the screen is sent from the FLEX 10K device to the monitor via five signals. Three VGA signals are red, green, and blue, while the other two signals are horizontal and vertical synchronization. Manipulating these signals allows images to be written to the monitor's screen. Figure 6.36 shows FLEX pin assignments to the VGA connector.

Mouse Connector. The mouse interface, is a six-pin mini-DIN connector that allows the FLEX 10K device to receive data from a PS/2 mouse or a PS/2 keyboard. The board provides power and ground to the attached mouse or keyboard. The FLEX 10K device outputs the DATA_CLOCK signal to the mouse and inputs the data signal from the mouse. Figure 6.37 shows the signal names and the mini-DIN and FLEX 10K pin connections.

Programming. EPF10K70 of the UP2 is programmed by the ByteBlaster II hardware connected to the JTAG terminal of the board. Board jumpers must be set as shown in Figure 6.38 in order to program this device.

6.3.4   Device Programming

Section 6.1.7 discussed device programming from Quartus II environment. Figure 6.15 shows Quartus II *programming* window that is used to program an EPM7128S device in JTAG programming mode. The device is connected to the LPT1 port through the ByteBlaster II programming cable.

In this section we show how the UP2 development board should be setup for its EPM7128S device to be programmed with the *anding* example of Section 6.1. We also show how this board is used for testing our simple demo example.

Programming Setup. To program the EPM7128S device of the UP2 board, connect the LPT side of ByteBlaster II to your computer and the JTAG end of it to the JTAG_IN of the UP2 board. Set the jumpers on the UP2 according to Figure 6.32 for programming the MAX device.

With this setup, a few seconds after clicking on the *Start Programming* icon of the *Programming* tool bar of Figure 6.16, the EPM7128S devices will be programmed with the *anding.pof* file that corresponds to our *anding* design.

Testing the Design. After performing the above steps, our *anding* project implemented on the MAX 7000S device of the UP2 board is ready to be tested. As discussed in Section 6.3.2, push-buttons and LEDs are not pre-assigned to EPM7128S pins. Figure 6.8 shows that input ports of our example design are assigned to device pins 54 and 56, and the output of our design is connected to pin 58. Figure 6.39 shows connections that need to be made from UP2 push-buttons and its *D2* LED to pins 54, 56, and 58 respectively.

With these connections we are ready to test the EPM7128S implementation of our *anding* design. Since the push-buttons are normally 1, the output of our design becomes 1 when the push-buttons are not pressed. However, since an LED is illuminated when a logic 0 is applied to it, the *D2* LED is initially off. By pressing one or both push-buttons, the output LED will be illuminated (AND output becomes 0). You will also notice that Segment *a* of MAX display digit 1 also illuminated when a push-button is pressed. This is because Pin 58 of EPM7128S is pre-assigned to this display segment (see Figure 6.31).

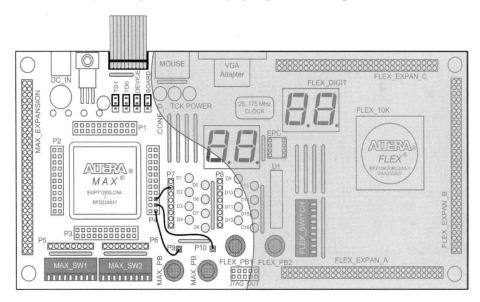

Figure 6.39 Programming MAX 7000S and Testing the *anding* Design

6.4 Summary

This chapter presented design entry, simulation and prototyping with tools that are provided by Altera for this purpose. We showed how an HDL simulator can be used for pre-synthesis simulation and then porting the tested design to Quartus II for synthesis and device programming. We also showed how a complete design can be directly entered into Quartus II for synthesis, post-synthesis simulation, timing analysis, and device programming. The design used here was a simple one; in the chapters that follow we will show implementation of more complex designs by running them through the design flow illustrated in this chapter.

7 Gate Level Combinational Design

This chapter shows the use of Quartus II for design of an iterative, hierarchical combinational circuit at the gate level. We will show steps necessary for entering a low level component and building upon that a larger combinational circuit. Testing and packaging the low-level component and putting it in a library for it to be used by its enclosing architecture will be illustrated. In the upper-level design that is presented, the use of bussing, multi-bit vectors, and use of functional simulation will be illustrated. The chapter shows programming the MAX device of UP2 and hardware testing of the design.

7.1 Element Design

The design we are doing in this chapter is an iterative comparator. This section shows project definition, design entry, simulation, and packaging of a 1-bit comparator that becomes the basic element of our n-bit comparator of the next section.

7.1.1 Project definition

The design of our 1-bit comparator is done as a new project of its own. This enables independent compilation and testing of this element. The project name is *bcompare* and is created in *chapter7* subdirectory of our Altera design directory.

7.1.2   Design Entry

Our design is a 1-bit cascadable comparator that has greater and equal outputs. As shown in Figure 7.1, our *bcompare* has *a* and *b* data inputs, *ei* (*equal-in*) and *gi* (*greater-in*) cascade inputs, and *eo* (*equal-out*) and *go* (*greater-out*) outputs. If *a* is greater than *b* and higher order bits are equal (*ei* is **1**), *go* becomes **1**. The *eo* becomes **1** when *a* and *b* are equal. This output will be used as the equal output of the comparator or the *ei* of the next higher order bit. The greater outputs will be ORed together to form the final greater output of the n-bit comparator built using this 1-bit design.

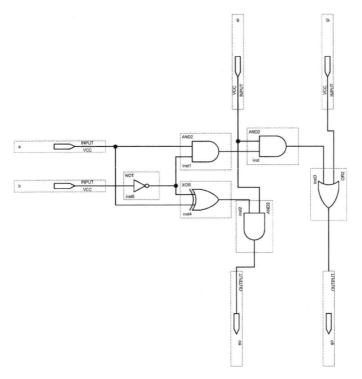

Figure 7.1 Cascadable 1-bit comparator

7.1.3   Functional Simulation

After completing the schematic entry of *bcompare*, it must be compiled. After a successful compilation, we will perform a functional simulation in order to verify our design. To begin this simulation, a new waveform file containing input data should be defined. For this purpose, in the *Applications* tool bar, select *New Vector Waveform File* and specify waveforms for the circuit inputs. From the main Quartus II tool bar, a waveform file can also be specified by selecting *File → New → Other Files → Vector Waveform File*.

Quartus II offers *functional* and *timing* simulation modes. In the *functional* mode, PLD delays are not considered and zero-delay simulation is performed, while the *timing* mode performs full timing simulation. The simulation mode is selected in the *Settings* window that can be reached from the *Assignments* pull-down menu. Among other simulation parameters, this window, shown in Figure 7.2, allows setting of the simulation mode.

Before running functional simulation, a netlist must be generated by running *Generate Functional Simulation Netlist* from the *Processing* menu. Figure 7.3 shows the functional simulation run of *bcompare*. Usually, after this simulation run and verifying the functionality of your design, you should run the simulation in the *timing* mode to make sure your applied input waveforms do not violate the worst-case delay of your circuit.

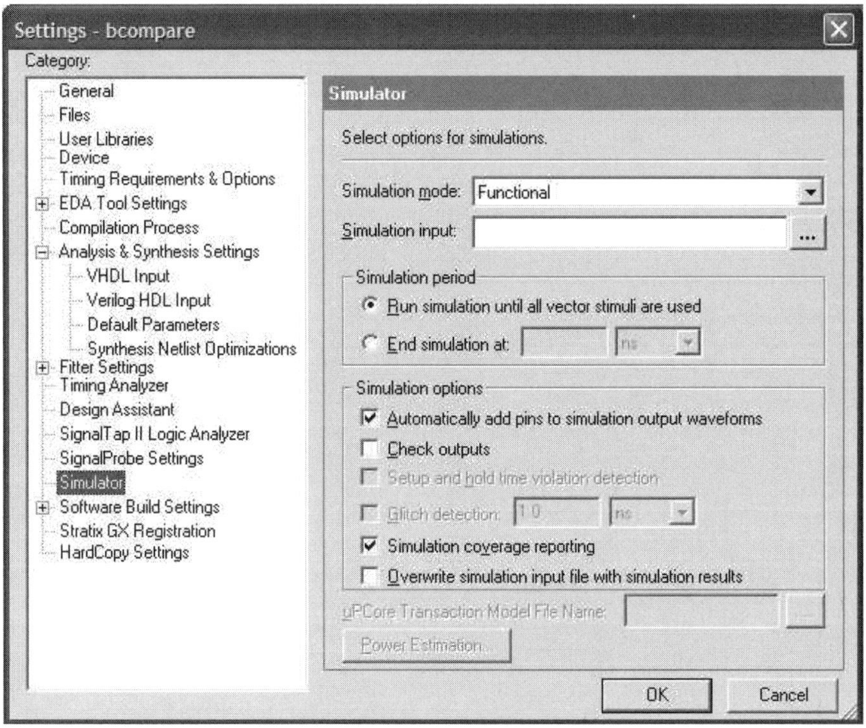

Figure 7.2 Simulator Settings in the *Settings* window

7.1.4 Packaging a Design

We can always cut and paste a design as many times as we want into a larger design that uses multiple instantiations of the lower-level design. A better alternative is to make a symbol for the design and use it as a package in an upper-level design. We package a general purpose design that is tested. The package will be placed in a library so that it becomes accessible to other designs.

bcompare Simulation Report
Simulation Report
- Legal Notice
- Flow Summary
- Flow Settings
- Simulator
  - Simulator Summa
  - Simulator Setting
  - Simulation Wave
  - Simulator INI Usɛ
  - Simulator Messag

Simulation Waveforms

Master Time Bar: 9.85 ns Pointer: 9.28 ns Interval: -570 ps Start: End:

Name	Value at 9.85 ns
a	B 0
b	B 0
ei	B 1
gi	B 0
eo	B 1
go	B 0

Figure 7.3 Functional Simulation of bcompare

Quartus II makes a default symbol for a design. For doing this, while the schematic window of *bcompare* is active, from the main Quartus II window, select *File → Create/Update → Create Symbol File for Current File*. The created symbol has a default shape that can be decorated by the user using the symbol editor of Quartus II.

To view and possibly edit the created symbol, go to the *File* menu select *Open* and open file *bcompare.bsf*. This opens in the symbol editor. In the symbol editor, the *Block and Symbol Editors* tool bar is used for editing a symbol. Figure 7.4 shows the default *bcompare* symbol with some user editing.

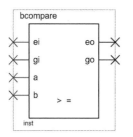

Figure 7.4 Symbol for bcompare

We close the *bcompare* project and start our 4-bit comparator project that uses this design.

7.2   Iterative Structures

In this section, the design of the 1-bit comparator of the previous section is used in an iterative fashion to form a 4-bit comparator.

7.2.1 Project Definition

We begin a new project in the *chapter7* directory that is the same directory as that of the *bcompare* project. The new project is named *icompare*. Because of the shared directory, files of *bcompare* can be used by *icompare*.

7.2.2 Design Entry

As with the *bcompare*, we use schematic entry for the *icompare* project. In the *Symbol* window that opens for entering design components, click on the little "+" next to *Project* to open this folder. In this folder all symbols defined in the subdirectory of this project will be shown. Since *bcompare* was defined in this project's subdirectory, it is accessible to *icompare*. Click on the *Repeat-insert mode* so that you can place multiple symbols in your design without having to come back to the *Symbol* window. Figure 7.5 shows the *Symbol* window in which *bcompare* is selected.

In the block diagram editor of *icompare*, place *bcompare* instances in a row. The left-most instance becomes the most-significant and the right-most is the least-significant bit. Double-click on the instance names to name them according their significance (see Figure 7.6).

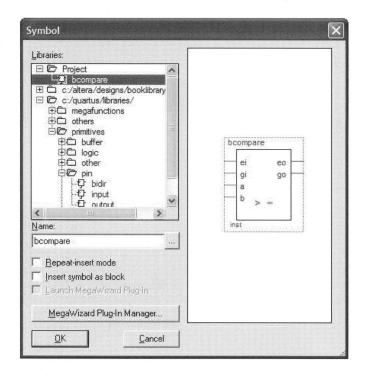

Figure 7.5 Placing Multiple Instances of *bcompare*

I/O Ports. For placing the I/O ports, we use *pin* primitive symbols from the Quartus II library of primitives. After these symbols are placed they must be named to represent input and output ports of the *icompare*. The 4-bit data ports are named *a[3..0]* and *b[3..0]*, and the output ports are named *equal* and *greater*. Naming an input port as *a[3..0]*, implicitly declares *a[3]*, *a[2]*, *a[1]*, and *a[0]*.

Busses. We use two 4-bit buses to connect our input ports to the 1-bit data inputs of our 1-bit comparators. To place a bus, click on the *Orthogonal Bus Tool* icon on the *Block and Symbol Editors* tool bar. Name these busses according to ports they are connecting to. Naming a bus is done by selecting the bus and setting its name in its *Bus Properties* window.

For connecting wires to these busses, start from the node of the component you are connecting to, and using the *Orthogonal Node Tool*, draw a wire to the 4-bit bus. For specifying which line of the bus a wire is connected to, select the wire and in its *Node Properties* window set its name to the name of the bus indexed by the bus-line it is connecting to. For example, naming a wire as *a[2]* connects the line to bit 2 of bus *a[3..0]*.

If name association is used, specific connections are not necessary, except for cosmetics reasons. The complete wiring of *icompare* based on *bcompare* is shown in Figure 7.6.

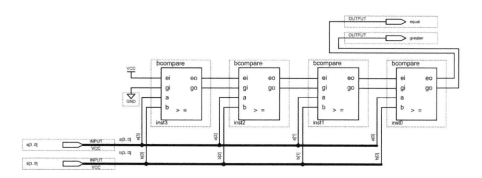

Figure 7.6 Complete *icompare* Block Diagram

Constants. Our design uses a constant **1** for its *equal-input* and a **0** for its *greater-input*. As shown in Figure 7.6, we use VCC and GND primitives for these constants.

7.2.3 Compilation

After completing the design entry phase, the design must be compiled. Compilation reports provide information on the utilization of the chip's hardware (macrocells for CPLDs, LEs for FPGAs), timing parameters, pins used, and routing and placement information.

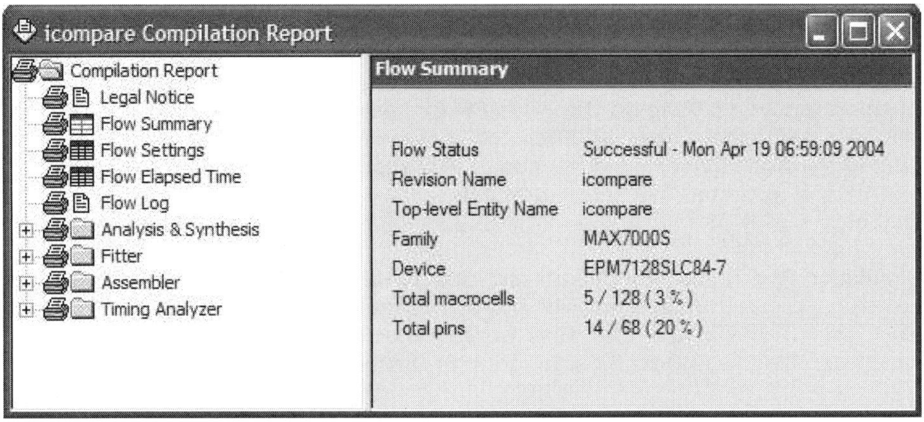

Figure 7.7 Compilation Summary

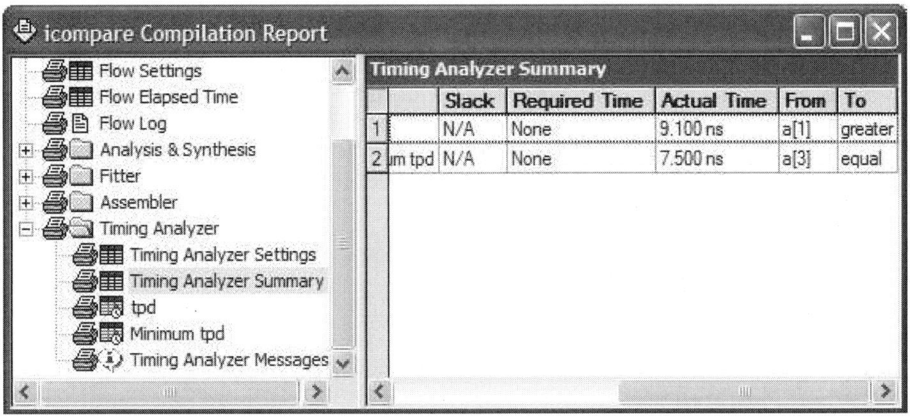

Figure 7.8 Timing Analyzer Summary

The *Flow Summary* of Figure 7.7 shows that the *icompare* design uses 5 macrocells of a MAX 7000S device. The *Timing Analyzer Summary* of Figure 7.8 indicates that the worst-case delay is 9.1 ns and occurs between the *a[1]* input and the *greater* output. Note that a different pin assignment can change this timing.

Compilation readies a design for device programming and simulation. After a design is compiled, the *Node Finder* utility will be able to look up design's nodes for pin-assignment and for waveform definition.

7.2.4 Simulation

In a waveform file, we should specify values for the *a[3..0]* and *b[3..0]* in order to apply to our *icompare* design and test it.

Selecting Input Vectors. To place the input vectors in the *Waveform* window, double-click in the signal name area and the *Insert Node or Bus* window opens. In this window, clicking on the *Node Finder* opens the *Node Finder* window. In this window, select "*Pins: all*" filter and click on *Start*. This will place all design pins in the *Node Finder* windows and allows users to select signals for assigning test values to them. We select *a* and *b* vectors to be placed in our *Waveform* window.

Simulation End Time. The default end-time for simulation is 200 ns. To extend our simulation run, click on *Edit* and in the pull-down menu that opens select *End Time ...* and in the *End Time* window that opens, enter a new simulation end-time. We use 600 ns for simulation of *icompare*.

Applying Test Data. We now show the procedure for assigning test vectors to inputs of *icompare*. In the *Waveform* window, double-click in the waveform area in front of the name of the bus or node for which a value is being assigned. This will select the bus' entire waveform interval for which we can specify test values.   Right-click on the selected area and in the menu that opens select *Value*. This will bring up a *Value* menu in which various schemes of applying test data to nodes or busses can be specified.

In our design, we use *Count Value* for *a* and *b* busses. In the *Count Value* window we can specify count values for our busses. We use a binary counter that increments every 71 ns for the *a* vector and a Gray code counter that increments every 113 ns for *b*. Figure 7.9 shows the *Count Value* window for *b*.

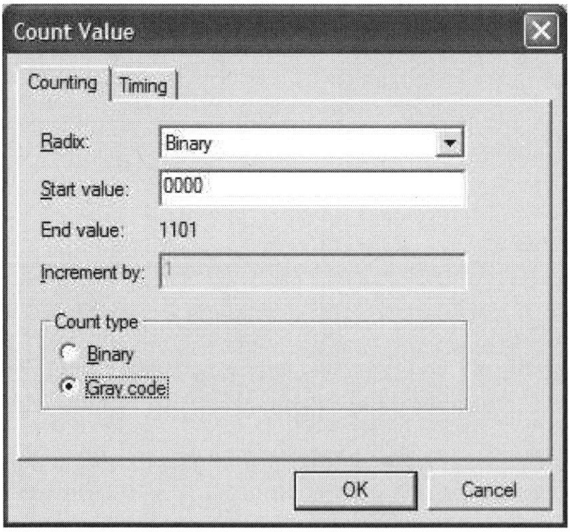

Figure 7.9 Gray Code Counter for *b[3..0]*

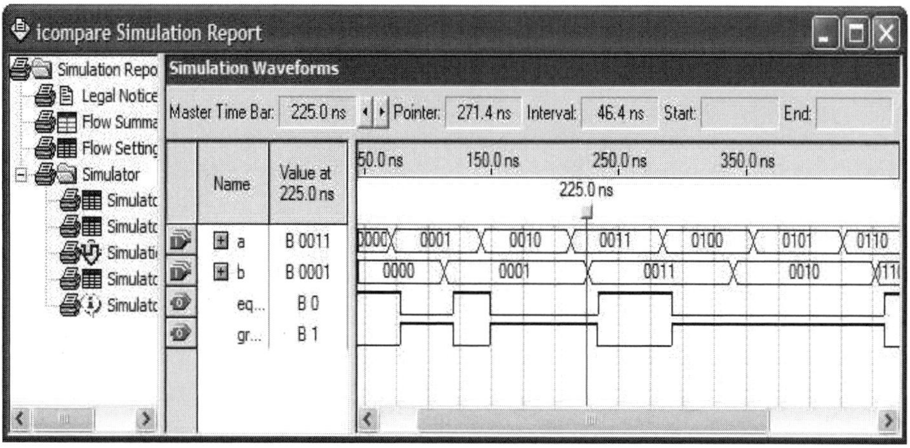

Figure 7.10 Simulation Result of *icompare*

In digital circuit simulation with timing, if input vectors change at the same time or if they change before the circuit has a chance to stabilize, it becomes difficult to trace the inputs and analyze the response of the circuit. In our test data, we have chosen 71 and 113 prime numbers to eliminate the chance of inputs changing at the same time. Because these numbers are far apart, the circuit will have a chance to stabilize before a new test vector is applied to it. Note that according to the compilation report of Figure 7.8, the worst-case delay of our design is 9.1 ns. Resulting waveform of the simulation run of *icompare* in the *timing* mode is shown in Figure 7.10.

7.2.5   Device Programming

After a successful simulation, we are now ready for device programming. We can either program our device with the pin-assignment done by Quartus II, or assign our own pins. In this design we do the latter.

Pin Assignment. In the Quartus II main window, clicking on the *Assignments* tab, brings up a pull-down menu in which selecting *Assign Pins* activates the window corresponding to this selection. A portion of this window is shown in Figure 7.11. In this window, a pin number is selected and a *Pin name* from our design is assigned to it. Since our design is already compiled, for looking for our design's pins we can use the *Node Finder* utility of Quartus II.

The *Node Finder* is activated by clicking on the three dots next to the *Pin name* box in the *Assign Pins* window. In our design, a[3], a[2], a[1], a[0], b[3], b[2], b[1], b[0], equal, and greater nodes are assigned to pins 24, 25, 27, 28, 29, 30, 31, 33, 37, and 39 respectively.

In assigning pins, make sure reserved pins, such as those of the JTAG, are not used. For example since pin 23 is used for TMS, it cannot be used as a regular I/O pin of our design.

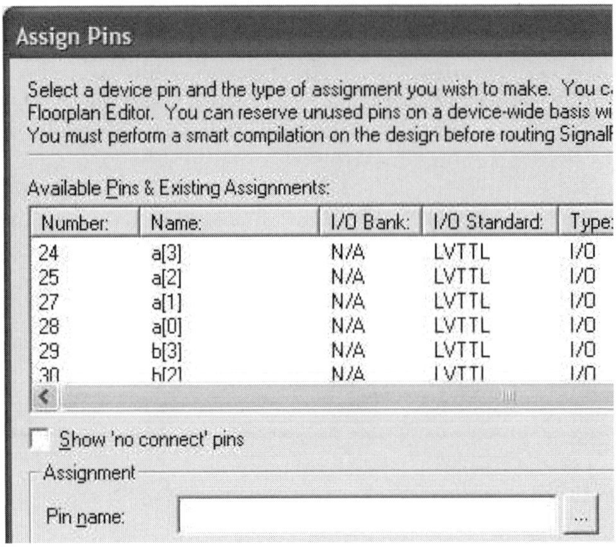

Figure 7.11 *Assign Pins* **Window (Partial View)**

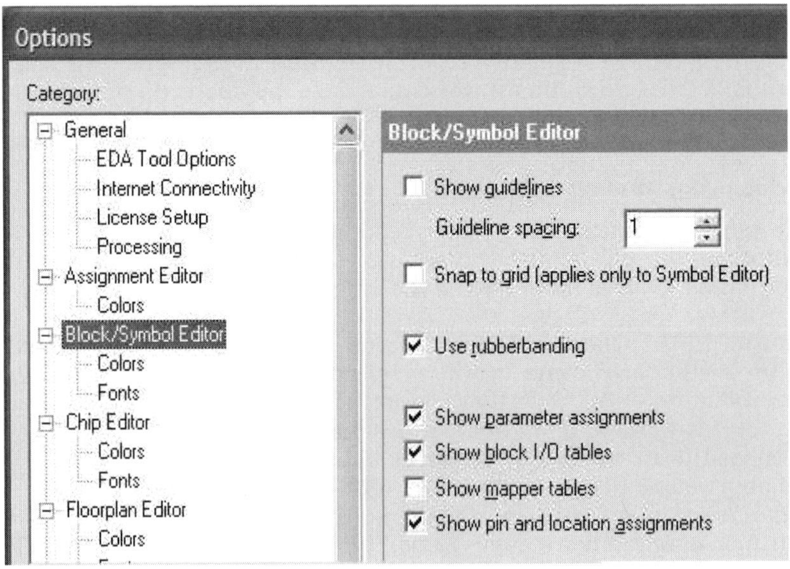

Figure 7.12 *Options* **Window (Partial View)**

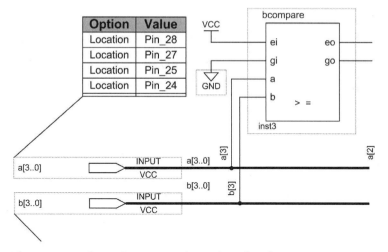

Option	Value
Location	Pin_28
Location	Pin_27
Location	Pin_25
Location	Pin_24

Figure 7.13 Pin Assignments Shown for *a[3..0]*

When pin assignments are complete, the design must be recompiled for the pin assignments to become effective in programming the device.

You can set an option to view pin assignments in the schematic window. For this purpose, click on the *Tools* tab and select *Options ....* This brings up the *Options* window (Figure 7.12) in which various categories of options are shown. Select the *Block/Symbol Editor* category, and in this category click on *General.* In the selection area that opens on the right-hand side part of this window, check *Show pin and location assignments.* When this is set, tabs next to ports of your design in its schematic window show pin numbers assigned to the I/O ports. Figure 7.13 shows a portion of schematic of *icompare* in which pin assignments for the *a[3..0]* port of this design are displayed.

Programming UP2. To start programming the MAX device of UP2, we connect the UP2 board to the printer port using ByteBlaster II and configure UP2 for programming this device (see Section 6.3.4). Then, in Quartus II, we run the *Programmer* utility to program our device. Make sure pins programmed as output are not connected to push-buttons or switches.

7.3   Testing the Design

With the programming discussed above, we can test our implementation of *icompare* on the UP2 board. Since, switch and LED connections to the MAX device are not pre-assigned in UP2, hard-wire connections should be made. We connect MAX_SW1 switch set to device pins 24, 25, 27, 28, 29, 30, 31 and 33, and D1 and D2 LEDs to pins 36 and 38. See Table 6.1 for exact positions of device pins in the prototyping headers surrounding EPM7128S.

Our design is tested by setting UP2 switches and observing the LEDs. Switches are at logic 1 in the up position, and the LEDs are illuminated when logic 0 is applied to them.

7.4 Summary

This chapter presented design and implementation of an iterative combinational circuit. The procedure we used for designing this circuit and entering it in Quartus II is typical of any large iterative circuit, such as ALUs. Using Quartus II, we showed steps for design entry and device programming on the UP2 board. Features of Quartus II that were not discussed in the previous chapter were discussed here. In the chapters that follow, only those features of Quartus II that we are seeing for the first time will be discussed.

8 Designing Library Components

This chapter shows design of small components that we will use in the chapters that follow as library components. The purpose is to show various ways a design can be implemented and at the same time generate a reusable library. The library components will be individually tested and symbols will be created for them. In order to test these components, each will be created as a complete project and they will be individually tested. Sequential and combinational components will be designed, and for their design, the use of primitives, megafunctions and Verilog for description of functions will be illustrated.

8.1 Library Organization

Components that we are discussing in this chapter will be placed in a directory called *BookLibrary*. Each component will have its own complete project in this directory. This way, we will be able to test each component by simulation and / or by device programming. For testing our library components we use the MAX 7000S device. Obviously, these components can be used in a design using any programmable device as long as they fit in the device.

8.2 Switch Debouncing – Schematic Entry

This section describes hardware for creating clean filtered pulses from mechanical UP2 pushbuttons. In design of this hardware we will use schematic entry at the gate and functional levels.

Pushbuttons on the UP2 board are mechanical switches and are not debounced. This means that when you press a pushbutton, it makes several

contacts before it stabilizes. The result is that when you press a pushbutton
that is **1** in the normal position, its output changes several times between logic
0 and **1** before it becomes **0**, and when you release it, it again switches several
times between these logic values before it becomes **1**. Figure 8.1 shows a
pushbutton contact bounce.

The problem described above causes no problems in combinational circuits
if you give enough time for all changes to propagate before reading the switch's
output. However, in sequential circuits with a fast clock, each of the bounces
between **0** and **1** logic values may be regarded as an actual logic value. For
example, for a counter with a fast clock for which a mechanical pushbutton is
used as a count input, pressing the pushbutton may cause several counts.

Figure 8.1 Contact Bounce in a Pushbutton

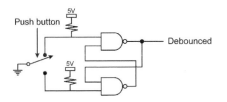

Figure 8.2 Debouncing a Single-Pole Double-Throw (SPDT) Switch

A Single-Pole Double-Throw (SPDT) mechanical switch such as that shown
in Figure 8.2 can easily be debounced by an SR-latch also shown in this figure.
However, UP2 pushbuttons are Single-Pole Single-Throw (SPST) and their only
available terminals are those that connect to logic **1** or logic **0**, as shown in
Figure 6.33.

Debouncing UP2 pushbuttons requires a slow clock to sample the switch
output before and after it is pressed or released. The clock should be slow
enough to bypass all the transitional changes that occur on its output terminal.
This section shows generation of a switch debouncer and its necessary clock.
For the design of the former part we use schematic entry at the gate level, and
for the latter part we use schematic entry using Quartus II megafunctions.

8.2.1   Debouncer – Gate Level Entry

The *debouncer* project is created in the *BookLibrary*. We use schematic entry at
the gate level for this design. The design is entered in Quartus II and is tested
on the EPM7128S device of UP2.

The design, shown in Figure 8.3, has two inputs *Switch* and *SlowClock*.
One of the flip-flops used here is triggered on the rising edge of *SlowClock* and
the other is triggered on the falling edge of this clock. Since the output of this
circuit is generated by ANDing the two flip-flop outputs, both flip-flops must see

logic **1** on their inputs before the output of the circuit becomes **1**. This means that the pushbutton connected to the *Switch* input of this circuit must stay high for the entire duration of the slow clock for the circuit output to become **1**.

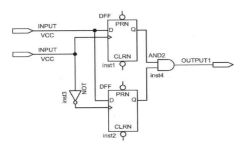

Figure 8.3 Schematic of the *Debouncer*

This design is entered in the Quartus II environment using its block editor. Flip-flops used here are part of the Quartus II library of primitives. These components are categorized in this library under *primitives/storage*. The specific flip-flop used is *dff*.

```
Debouncer
                   DebouncedSwitch
x —| Switch                          —x
x —| SlowClock

 inst
```

Figure 8.4 Default Symbol Created by Quartus II for *Debouncer*

After entering the schematic of this design, a symbol (shown in Figure 8.4) is created for it by using the Quartus II utility for generating default symbols. This part of the hardware for debouncing pushbuttons uses 3 of the 128 macrocells of the MAX 7000S device.

This part of our design can be simulated, but the real test of this circuit is using it with UP2 pushbuttons. This requires the use of a slow clock that will be created next.

8.2.2   Slow Clock – Using Megafunctions

The frequency of the UP2 clock is 25.175 MHz. Obviously this is too fast for filtering transitions in pushbuttons. Dividing this clock by 2^{21} produces a 12 Hz clock that will be more adequate for filtering slow mechanical transitions. We use a 21 bit counter for dividing the UP2 on-board clock. The Quartus II project for this purpose is called *Divider21* and is created in the *BookLibrary* directory. We demonstrate the use Altera megafunctions for the generation of this circuit.

The design shown in Figure 8.5 uses a 21-bit up-counter. The input is the fast clock and bit 20 of the counter output is the slow clock. The core of this counter is *Divider21c* that is made by configuring a Quartus II megafunction.

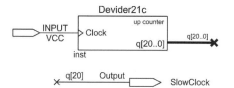

Figure 8.5 Slow Clock Generator

Megafunctions. Just like gates and flip-flops, megafunctions are part of the Quartus II library of components. Unlike gates and flip-flop primitives that are only available with predefined features and inputs, megafunctions are configurable. For example, an OR megafunction can be configured to become an array of n-input OR gates. A counter megafunction can be configured as an up- or down-counter with any number of bits with various forms of load, reset and preset control inputs.

In general, megafunctions are frequently used general purpose digital parts that can be customized according to specific applications. In a way, megafunctions replace the older 7400 series of parts that are available in many technologies for board level designs. The 7400 series packages cover a wide range of functions, but because they are actual physical parts, they only have a limited configurability. Altera megafunctions also cover a wide range of functions. They are described in a hardware description language and because of this, they are far more flexible than the 7400 series that are physical parts.

Megafunctions are available in five categories: *arithmetic*, *embedded_logic*, *gates*, *IO* and *storage*. The *arithmetic* megafunctions cover various forms of adders, counters, and other general purpose arithmetic functions. The *storage* category covers memories, registers, RAMs and ROMs.

Quartus II utility for configuring megafunctions is *MegaWizard Plug-In Manager*. When this utility is invoked, in a series of windows it asks users to specify and configure the megafunction that they have chosen. When done, it generates a schematic symbol for the configured part and generates an HDL design file that corresponds to the symbol. The symbol can be placed on the block editor and used with other configured megafunctions or primitives for completing a design.

To access a megafunction, go through the same process as for placing a primitive in your schematic. When the *Symbol* window appears, in the list of libraries select the standard Quartus II library (i.e., \quartus\libraries\) and open the *megafunctions* folder in this library. In what follows we show how the *Divider21c* counter of Figure 8.5 is generated.

Frequency Divider. To enter the megafunction counter in the schematic of *Divider21* project, while the schematic entry window (*Block and Symbol Editors*) is open, select the *Symbol Tool* from the corresponding tool bar. This opens the *Symbol* window shown in Figure 8.6. Open this library and in the *arithmetic* category, select *lpm_counter*. After clicking *OK*, a series of windows will appear that allow you to configure your counter.

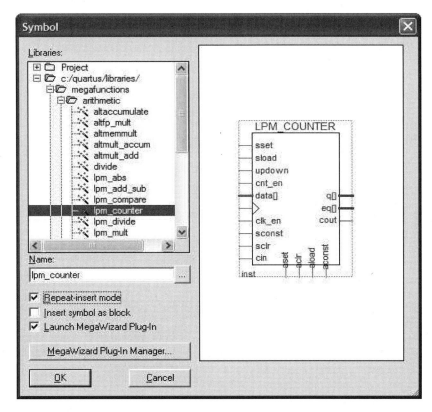

Figure 8.6 Megafunctions Library

In the first window, the HDL used for specifying this megafunction and the name you want to give it is defined. The language used can be any of the three choices shown. Use any language you are most comfortable with. For the name of the megafunction we use *Divider21c*. The next three windows allow you to specify count direction (up or down) number of bits, count sequence (binary or Modulus), enabling mechanism, and set or resetting mechanism (synchronous, asynchronous, etc). Figure 8.7 shows one of these three windows. The last window tells you the files that are generated and added to your project when this mega function is generated. One of the files created is *Divider21c.bsf* that represents the symbol that corresponds to your configured counter. When configuration of a megafunction is complete, this symbol is placed on your schematic (see Figure 8.5).

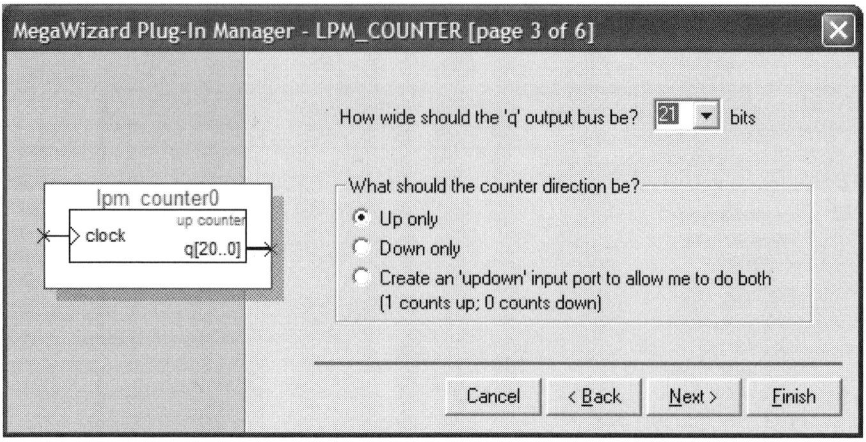

Figure 8.7 Counter Megafunction Configuration Window

After placing *Divider21c* on the schematic, the design of *Divider21* frequency divider will be complete by connecting the input *Clock* to the clock input of *Divider21c* and the *SlowClock* output to *q[20]*. As shown in Figure 8.5, connection to *q[20]* port of *Divider21c* is done by connecting a bus to the *q* output of this component, assigning a name to it and using the indexed name for driving the *SlowClock* output.

This completes the design of the frequency divider. In order for this design to be usable in other designs, a symbol is generated for it. This symbol will be used in the complete design of the switch debouncing hardware.

8.2.3 A Debounced Switch – Using Completed Parts

Finally by putting together the hardware of Figure 8.3 with that of Figure 8.5 hardware for debouncing a switch is generated. For this hardware we generate the *Debounced* project in the *BookLibrary* directory, and in its schematic we use symbols for *Debouncer* (Figure 8.4) and *Divider21*.

For placing these symbols in the schematic of *Debounced* project, click on the *Symbol Tool* of the *Block and Symbol Editors* toolbar. When the *Symbol* window opens, in the *Libraries* hierarchy select *Project* (first item in Figure 8.6). This points to symbols created in the directory of our present project. Since the *Debouncer* and *Divider21* projects use the *BookLibrary* directory, their symbols are available in the *Project* directory.

When the *Project* hierarchy opens, select *Debouncer* and *Divider21* symbols and place them on the schematic of the *Debounced* project. The complete schematic diagram of this design is shown in Figure 8.8. The *SlowClock* signal feeds the input of the *Debouncer*; in addition it is pulled out as an output of the *Debounced* hardware. This way, the *SlowClock* output can be shared among multiple *Debouncer* components.

A symbol, shown in Figure 8.9, is created for this design so that it can be used in designs requiring a debounced pushbutton.

For testing our project, it is compiled and it is programmed into the MAX 7000S device of UP2. The *FastClock* input is put on pin number 83 that is the global lock of this chip. The *Switch* input is wired to a pushbutton, and the *CleanSwitch* and *SlowClock* outputs are put on two of the MAX LEDs. The functionality of this circuit is tested by pressing a pushbutton and observing its output. Note that if glitches transmit to the output, because of their short duration, we will not be able to see them on the output LED.

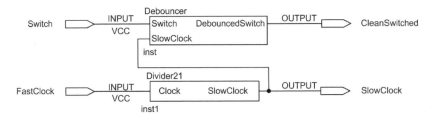

Figure 8.8 Schematic of the *Debounced* Project

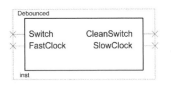

Figure 8.9 *Debounced* Symbol

The complete hardware of the *Debounced* component uses 24 of the 128 macrocells of the MAX 7000S device. Because of this high gate count, it is recommended that for multiple switches, only one *Debounced* is used and the rest use the *Debouncer* of Section 8.2.1 that only uses 3 macrocells.

8.3   Single Pulser – Gate Level

Often, start pulse for a sequential circuit must be only one clock pulse duration. A problem with using pushbuttons for this purpose is that operation of such a switch by human is usually very slow, and the best we can do is to generate pulses of several milliseconds by pushing a push button.

The *OnePulser* project of this section takes a clock and a long pulse as inputs and produces a single pulse of the duration of the clock period for every time the long pulse becomes **1**. The output pulse is synchronous with the clock. The long-pulse input connects to a debounced switch, and the clock input of this circuit connects to the main clock signal of the sequential circuit using the start pulse. The design is done using a block diagram using primitives shown in Figure 8.10.

This design is done by a 2-bit shift register. As shown in Figure 8.10, when *LongPulse* is **0** on the rising edge of the *ClockPulse* at time t_i and **1** on the rising edge of *ClockPulse* at t_{i+1}, the *OnePulse* output becomes **1** after the edge at t_{i+1} and remain **1** until the next edge that a **1** is shifted into the shift-register. The output waveform of this circuit that is actually a **01** detector is shown in Figure 8.11.

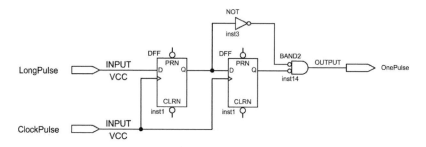

Figure 8.10 The *OnePulser* Circuit

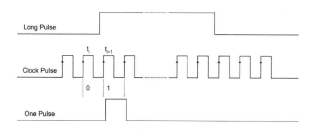

Figure 8.11 *OnePulser* Output Waveform

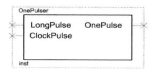

Figure 8.12 *OnePulser* Symbol

The *OnePulser* circuit is tested by instantiating a copy of the *Debounced* circuit (Figure 8.8) in its schematic and driving its inputs. The *CleanSwitch* and *SlowClock* outputs of *Debounced* connect to the *LongPulse* and *ClockPulse* inputs of *OnePulser*, respectively. The *Switch* and *FastClock* inputs of *Debounced* connect to a pushbutton and the system clock respectively, and the *OnePulse* output is displayed on an LED on the UP2 board. By pressing the input pushbutton, a short blink on the output LED indicates the correct operation of this circuit.

The *OnePulser* uses 3 of the 128 macrocells of the MAX device on UP2.

8.4   Debouncing Two Pushbuttons – Using Completed Parts

We generate a project and a design called *Pulser2* to debounce both MAX pushbuttons of the UP2 board. This circuit uses two copies on the *Debouncer* of Figure 8.4 and the *Divider21* of Figure 8.5. This design is shown in Figure 8.13.

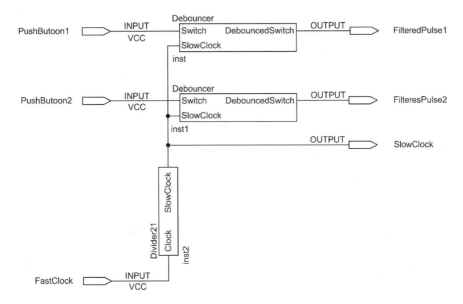

Figure 8.13 *Pulser2* **Schematic**

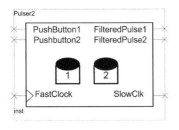

Figure 8.14 *Pulser2* **Symbol**

The *Pulser2* symbol is shown in Figure 8.14. For generation of this symbol we first use Quartus II symbol generation utility to generate a default symbol. Using the symbol editor tools (on the *Block and Symbol Editor* toolbar), the default symbol is edited to look as shown in this figure. Tools used from this toolbar are *Text, Rectangle, Oval,* and *Line* tools.

8.5   Hexadecimal Display – Using Verilog

The next library component described in this chapter is a hexadecimal display driver. As with the other components, we will generate a design file and a symbol for this circuit. This circuit takes a 4-bit HEX input and generates Seven Segment Display (SSD) code that corresponds to the input data.

Alternative methods of design entry that exist for creating this design include gate-level schematic entry, table-driven ROM specification, use of megafunctions, use of standard 7400 parts (7400 parts are available in *Libraries\others\maxplus2*), and using an HDL like Verilog. We have chosen the latter method, since it is easy to describe, uses fewer device cells than some other methods, and is adaptable to both MAX and FLEX. The entry method used here starts from the block diagram editor of Quartus II, and enables the use of Verilog for describing a block used in the schematic diagram.

The project used for the display adapter is called *DisplayHEX* and is created in the *BookLibrary* directory. Once this project is defined, we use the *New Block Diagram/Schematic File* tool of the *Applications* toolbar to open a new schematic file. In the schematic, the *HexDecoder* block will be defined to include the Verilog code for the core of our project.

8.5.1   Block Specification

When a new schematic file opens, its corresponding toolbar (*Block and Symbol Editors*) becomes available, and the schematic window is initially blank. Select the *Block* tool from this toolbar to place a blank block on the schematic window. The size of this block is not important at this point and can be adjusted once more details are known about it. Figure 8.15 shows a block that needs to be configured. This becomes our *HexDecoder*.

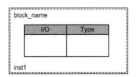

Figure 8.15 An Empty *Block*

8.5.2   Block Properties

The next step in defining a block for inclusion of a design description in Verilog is to specify its name and ports. This is done by specifying block properties. Right-click on the block of Figure 8.15, and in the pull-down menu that opens select *Block Properties*. In the *General* tab of the window of Figure 8.16 the block name (*HexDecoder*) is entered and in its *I/Os* tab its input and output ports are specified.

The 4-bit input of *HexDecoder* is *HEXin[3..0]* and its 7-bit output is *SSDout[6..0]*. The next step is entering the Verilog code of *HexDecoder*.

Block Properties

General | I/Os | Parameters | Format

I/O

Name: [   ] ▼   Add

Type: [OUTPUT ▼]   Delete

Existing block I/Os:

Name	Direction
HEXin[3..0]	INPUT
SSDout[6..0]	OUTPUT

OK   Cancel

Figure 8.16 Block Properties

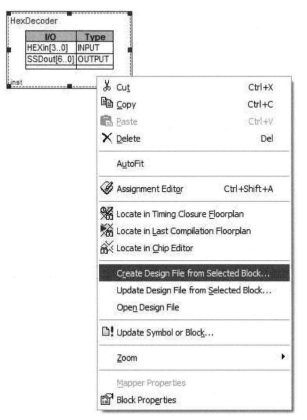

HexDecoder

I/O	Type
HEXin[3..0]	INPUT
SSDout[6..0]	OUTPUT

inst

✂ Cut	Ctrl+X
🖹 Copy	Ctrl+C
📋 Paste	Ctrl+V
✕ Delete	Del
AutoFit	
🖊 Assignment Editor	Ctrl+Shift+A
🖼 Locate in Timing Closure Floorplan	
🖼 Locate in Last Compilation Floorplan	
🖼 Locate in Chip Editor	
Create Design File from Selected Block...	
Update Design File from Selected Block...	
Open Design File	
🖹! Update Symbol or Block...	
Zoom	▶
Mapper Properties	
🖺 Block Properties	

Figure 8.17 Creating a Verilog HDL Design File

8.5.3   Block Verilog Code

With the block properties defined above, Quartus II can generate a Verilog template file for entering the Verilog code of *HexDecoder*. In the schematic of *DisplayHEX*, right-click on the block symbol of *HexDecoder* and in the pull-down menu that opens (Figure 8.17) select *Create Design File from Selected Block* .... This generates a Verilog template that contains declarations and I/O ports of the *HexDecoder* **module**.

The complete Verilog code of *HexDecoder* is shown in Figure 8.18. The *HexDecoder.v* file in the *BookLibrary* contains this code. What is shown here in bold is the part of the code that we have entered for the description of our design. The rest of the code has been generated automatically by Quartus II. The description of *HexDecoder* is now complete.

```
// Generated by Quartus II Version 3.0 (Build Build 199 06/26/2003)
// Created on Thu Mar 11 02:53:38 2004

// Module Declaration
module HexDecoder
(
     // {{ALTERA_ARGS_BEGIN}} DO NOT REMOVE THIS LINE!
     HEXin, SSDout
     // {{ALTERA_ARGS_END}} DO NOT REMOVE THIS LINE!
);
// Port Declaration

     // {{ALTERA_IO_BEGIN}} DO NOT REMOVE THIS LINE!
     input [3:0] HEXin;
     output [6:0] SSDout;
     // {{ALTERA_IO_END}} DO NOT REMOVE THIS LINE!

     assign  SSDout =
     HEXin == 4'b0000 ? 7'b0000001 :
     HEXin == 4'b0001 ? 7'b1001111 :
     HEXin == 4'b0010 ? 7'b0010010 :
     HEXin == 4'b0011 ? 7'b0000110 :
     HEXin == 4'b0100 ? 7'b1001100 :
     HEXin == 4'b0101 ? 7'b0100100 :
     HEXin == 4'b0110 ? 7'b0100000 :
     HEXin == 4'b0111 ? 7'b0001111 :
     HEXin == 4'b1000 ? 7'b0000000 :
     HEXin == 4'b1001 ? 7'b0000100 :
     HEXin == 4'b1010 ? 7'b0001000 :
     HEXin == 4'b1011 ? 7'b1100000 :
     HEXin == 4'b1100 ? 7'b0110001 :
     HEXin == 4'b1101 ? 7'b1000010 :
     HEXin == 4'b1110 ? 7'b0110000 :
     HEXin == 4'b1111 ? 7'b0111000 :
                7'b1111111 ;
endmodule
```

Figure 8.18 *HexDecoder* **Verilog Code**

8.5.4   Connections to Block Ports

When a block is defined, it can be used like any symbol in a design file. The *HexDecoder* block in the *DisplayHEX* design file is completely defined and the next step is to wire its ports to other components of this design. Place an input pin symbol and an output pin symbol in *DisplayHEX* schematic.   Assign *HEX[3..0]* for the input pin name and *SSD[6..0]* for the output pin. This makes the input a 4-bit bus and the output of *DisplayHEX* a 7-bit bus.

Use the *Orthogonal Bus* tool to make connections from *HEX[3..0]* and *SSD[6..0]* pins to the sides of the *HexDecoder* block. I/O mapper symbols will be placed on the block boundaries where bus connections touch the *HexDecoder* block (see Figure 8.19).

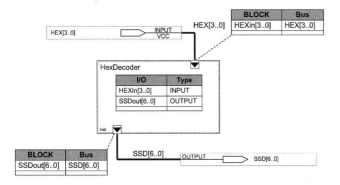

Figure 8.19 Connections of the *DisplayHEX* to *HexDecoder* Block

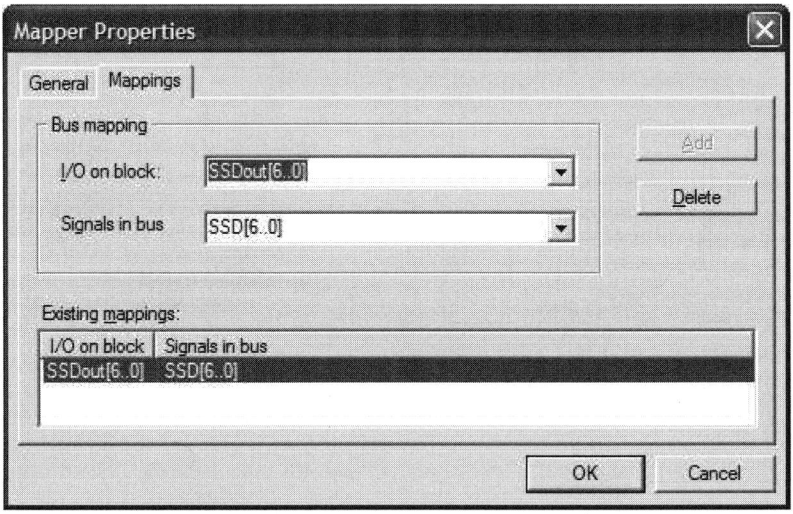

Figure 8.20 Block Port Mapper

The mappers shown in Figure 8.19 must be configured to map outside busses to the internal ports of *HexDecoder*. To configure a mapper, right-click on the mapper on the boundary of *HexDecoder* to bring up the *Mapper Properties* window, shown in Figure 8.20. In the *Mappings* tab of this window select an I/O on block and make a mapping between that and a signal in bus. Figure 8.20 shows a mapping made between *SSDout[6..0]* of *HexDecoder* and *SSD[6..0]* bus of *DisplayHEX*.

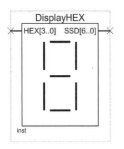

Figure 8.21 Symbol for *DisplayHEX*

8.5.5   Completing *DisplayHEX*

The schematic shown in Figure 8.19 is the complete design of *DisplayHEX* and is compiled in the *BookLibrary*. For it to be accessible by other designs, a symbol is generated that is shown in Figure 8.21. This design uses 5 of the 128 macrocells of the MAX device of UP2 board.

8.6   Summary

This chapter presented various ways designs can be generated in Quartus II. At the same time we presented several utility hardware structures. The structures presented are put in a library to be accessible by designs of the following chapters. On the use of Quartus II, this chapter showed definition and usage of megafunctions, defining and using HDL blocks, using existing components in a design, and editing and customizing component symbols. On the organizational side, this chapter showed how a library of parts could be generated and tested. Finally, from digital design point of view, this chapter showed small, but useful, parts that many designs can use.

9 Design Reuse

This chapter shows digital system design by reuse of already tested parts. For demonstration of this topic we use a simple counter circuit and take advantage of components designed in the previous chapter.

9.1 Design Description

The design we are using for demonstration of reuse of existing parts is a hexadecimal up/down counter with synchronous count enable and clear inputs. The counter counts from 0 to 15 or 15 down to 0 depending on its up or down mode of count.

For the design of the counter we use an Altera megafunction and configure it according to the above specification.

The counter uses the fast 25 MHz clock that is available on UP2. For control of the counter we use UP2 pushbuttons and switches. Components from *BookLibrary* for pushbutton debouncing and single pulse generation will be utilized in this project. The counter output will be displayed on one of the Seven Segment Displays of UP2, for which we will take advantage of the *DisplayHEX* library component.

9.2 Project Definition

The design project we use for our counter circuit is *HexCounter* and will be created in a directory with the same name.

We use the *New Project Wizard* for creating our *HexCounter* project. The procedure is the same as other projects we have discussed so far, except that we have to specify that this project uses design files from the *BookLibrary*.

By default, *quartus/libraries* and the current project's library (*Project*) are available to all designs. If more libraries are needed, they have to be explicitly specified during project creation, or in the *Settings* window.

During the project definition phase, we are asked to specify other design files used by this project. We will be allowed to specify *User Library Pathnames* for libraries that we are accessing. The window shown in Figure 9.1 is where user library path names are specified. We add the directory of our *BookLibrary* to the list of available libraries.

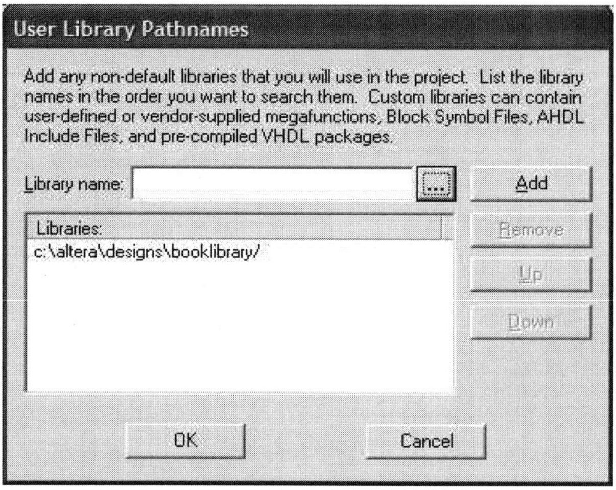

User Library Pathnames

Add any non-default libraries that you will use in the project. List the library names in the order you want to search them. Custom libraries can contain user-defined or vendor-supplied megafunctions, Block Symbol Files, AHDL Include Files, and pre-compiled VHDL packages.

Library name: [   ] [...]   Add

Libraries: Remove
c:\altera\designs\booklibrary/
                                                  Up

                                                  Down

        OK   Cancel

Figure 9.1 User Library Pathnames

9.3 Design Implementation

The *HexCounter* described here has a counter core and several interfaces. We use Altera's megafunctions for the core of the counter and our pre-defined library functions for its interfaces.

We begin this design by opening a schematic window in Quartus II. The counter core and its interfaces will be placed here and proper interconnections made between them.

9.3.1 Counter Core

The *lpm_counter* megafunction of Quartus II is available in the *quartus\libraries\megafunctions\arithmetic* directory of Quartus II libraries. This library component can be adjusted for size, count sequence, control inputs, and count direction. The original schematic of this counter is Figure 9.2.

When this megafunction is selected for being placed in a schematic, through a series of interactive windows, the *Megafunction Wizard* generates a counter according to the designer's specifications. For our design, we have communicated the following specifications to the *Megafunction Wizard*. As a result, a counter is built and its symbol is shown in Figure 9.3.

- Use Verilog HDL
- Output file name is *CounterCore*
- Output bus is 4 bits wide
- Create an up-down input to allow count-up and count-down
- Counter is a binary counter
- Provide a count enable input
- Provide synchronous clear

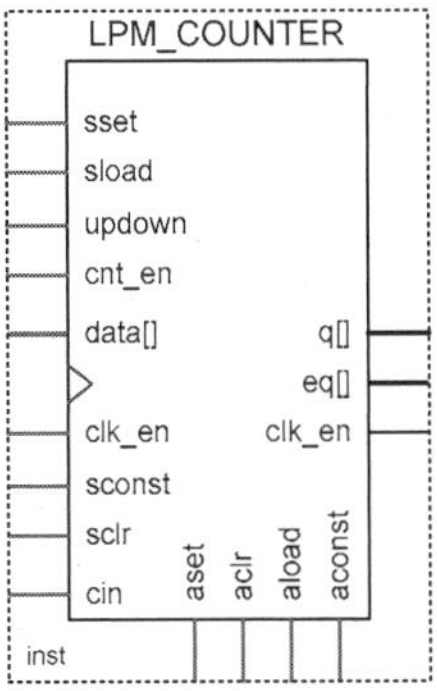

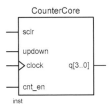

Figure 9.2 Quartus II Configurable Counter

Figure 9.3 Configured Counter Core

9.3.2   Counter Interfaces

The counter of Figure 9.3 has three control inputs, *sclr* for synchronous clear, *updown* for up-down control, and *cnt_en* for count enable. We use a UP2 switch for the up-down control and pushbuttons for the clear and count-enable inputs.

For debouncing the pushbuttons we use the *Pulser2* library component of the *BookLibrary*. Pulses generated by *Pulser2* are long and they are active for as long as a pushbutton is pressed. For debounced outputs of the pushbuttons, we use *OnePulser* library components to generate a single pulse coinciding with our circuit clock for every time a pushbutton is pressed.

The counter output is a 4-bit hexadecimal code. We use our *DisplayHEX* SSD driver to display the counter output on one of the UP2 displays.

The complete schematic of our *HexCounter* is shown in Figure 9.4. For connections between various components of this design we use the *Orthogonal Node* or the *Orthogonal Bus* tool. Connections from the *DisplayHEX* output to the output pins of our design are done by assigning the same signal names to connecting signals.

9.3.3   Pin Assignments

Our design project in this chapter is small enough to fit on MAX or FLEX devices, and it does not require any of the special features that are only available in FLEX devices. For example, using ROMs or RAMs are only available on certain Altera devices, and not on the CPLDs. In this project, we have chosen to implement this design on the MAX CPLD. In our case, the only difference between the FLEX and MAX is the different pin assignments.

Inputs of the *HexCounter* design are *UpDown*, *Clear*, *Count* and *SystemClock*. The outputs of this design are *SSD[6]* through *SSD[0]*. Since *UpDown* is a mode select input debouncing it is not necessary and we use one of UP2 switches. *Clear* and *Count* inputs connect to the pushbuttons. The *SystemClock* input of our circuit connects to the UP2 main clock that is directly connected to pin 83 of the MAX device.

Display outputs *SSD[6]* through *SSD[0]* connect to MAX Digit 1 according to pin assignment of Figure 6.31.

Table 9.1 shows *HexCounter* port connections to EPM7128S pins and UP2 resources. The first three connections shown in this table are not part of UP2 and must be made by inserting connecting wires.

Table 9.1   HexCounter Port Connections

HexCounter Ports	MAX 7000S Pins	UP2 Resources
UpDown	24	Wired to MAX Switch 1
Clear	56	Wired to MAX PB 2
Count	54	Wired to MAX PB 1
SystemClock	83	25 MHz Clock
SSD[6]	58	MAX Digit 1, Segment a
SSD[5]	60	MAX Digit 1, Segment b
SSD[4]	61	MAX Digit 1, Segment c
SSD[3]	63	MAX Digit 1, Segment d
SSD[2]	64	MAX Digit 1, Segment e
SSD[1]	65	MAX Digit 1, Segment f
SSD[0]	67	MAX Digit 1, Segment g

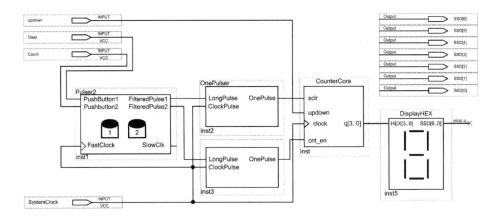

Figure 9.4 *HexCounter* **Complete Design**

9.3.4 RTL View

The RTL Viewer tool is a utility for viewing the post-synthesis schematic of the synthesized design, independent from the target device that is being programmed. This view of a design becomes available after compiling it. Figure 9.5 shows a portion of the RTL view of the *HexCounter*. In this view, by clicking on a component of the design, the RTL Viewer shows the details of that component. Figure 9.6 shows the RTL view after clicking on *CounterCore*. Note in this view, the use of flip-flops and primitive gates.

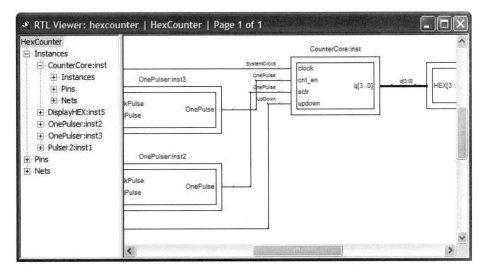

Figure 9.5 **Top-Level RTL View of** *HexCounter*

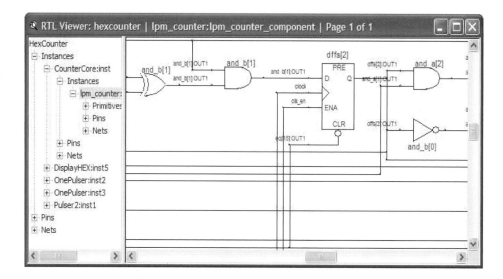

Figure 9.6 RTL View of *CounterCore* of *HexCounter*

9.3.5   Floorplan View

After compiling *HexCounter*, several report files are created that indicate the timing, cell usage, and macrocells used by this design.   The compilation summary indicates that a total of 40 macrocells out of 128 available macrocells are used by this design.

The floorplan view that is generated by the compiler shows macrocells used by various parts of our design.   Starting from the schematic of our design, we can locate macrocell(s) that correspond to a certain block of our design. This is done by right-clicking a block in the block diagram of a design, and selecting *Locate in Last Compilation Floorplan* in the pull-down menu that opens.

9.3.6   Device Programming

The *HexCounter.pof* file that is generated by the compiler is used by the programmer to program the MAX 7000S device of UP2. Because this device is EEPROM based, a design that is programmed in it remains there until overwritten by another.

9.4   Summary

We have shown implementation of a design using megafunctions from the standard Quartus II library and pre-tested components from a user library. No logic level design or Verilog coding was necessary for the implementation of *HexCounter*.

10 HDL Based Design

In this chapter we show a design done in Verilog and implemented on UP2. We describe a circuit at the behavioral level and test it with an HDL simulator. The tested design is then brought into Quartus II and incorporated into a complete design. The Quartus II compiler synthesizes our behavioral code and generates a device programming file that contains the synthesized parts as well as other components used in the complete design. The final design is used to program the MAX device of the UP2 board.

10.1 High Level Description and Simulation

The design that will be used in this chapter is a sequence detector. After describing the functionality of this circuit, we will describe it in Verilog, test our synthesizable Verilog design with an HDL simulator, and ready it for being used in programming a programmable device. This section describes the design of the sequence detector and the next section discusses its preparation for device programming.

10.1.1 State Machine Description

The core of the design that will be used to program the MAX device of UP2 is a sequence detector that detects a sequence of **1011** on its x serial input. When this sequence is detected, it generates a complete one clock duration pulse on its z output. The machine has a reset input (*rst*) that forces the machine into its initial or reset state. The state machine for implementing this design is a Moore machine, the state diagram of which is shown in Figure 10.1.

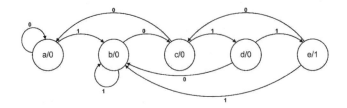

Figure 10.1 Moore 1011 State Diagram

```
`timescale 1ns/100ps

module moore1011 ( x, rst, clk, z );
  input x, rst, clk;
  output z;

  parameter [2:0] a= 0, b = 1, c = 2, d = 3, e = 4;
  reg [2:0] p_state, n_state;

  always @( p_state or x ) begin: combinational
    case ( p_state )
      a:
        if( x == 1 ) n_state = b;
        else n_state = a;
      b:
        if( x == 0 ) n_state = c;
        else n_state = b;
      c:
        if( x == 1 ) n_state = d;
        else n_state = a;
      d:
        if( x == 1 ) n_state = e;
        else n_state = c;
      e:
        if( x == 1 ) n_state = b;
        else n_state = c;
      default:
        n_state = a;
    endcase
  end

  assign z = (p_state == e);

  always @( posedge clk ) begin: sequential
    if( rst ) p_state = a;
    else p_state = n_state;
  end

endmodule
```

Figure 10.2 Moore Detector Verilog Code

10.1.2 Moore Machine Verilog Code

The synthesizable Verilog code of our Moore machine is shown in Figure 10.2. As shown, three concurrent statements describe transitions, output and clocking of this machine. The *combinational* **always** block describes the transitions shown in the state diagram of Figure 10.1, while the *sequential* **always** block describes the clocking of our machine. The other concurrent statement in this description is an **assign** statement that is used for assigning values to the z output of the circuit.

The description shown uses *p_state* for the present state of the machine and *n_state* for its next state. Because our machine has five states, parameter and **reg** declarations are three bits wide.

10.1.3 Moore Machine Verilog Testbench

The testbench of Figure 10.3 is used for testing the Verilog code of Figure 10.2. In this testbench a 19-bit wide buffer holds serial data that are applied to the x input of the circuit being tested. Two concurrent **always** blocks in this description produce the circuit clock and simultaneously rotate *buffer* bits out into x.

```
`timescale 1ns/100ps

module test_moore1011;
  reg x, rst, clk;
  wire z;
  reg [18:0] buffer;

  moore1011 uut( x, rst, clk, z );

  initial buffer = 19'b0001101101111001001;
  initial begin
    clk = 0; x = 0; rst = 1;
    #29 rst = 0;
    #500 $stop;
  end
  always @(posedge clk) #1 {x, buffer} = {buffer, x};
  always #5 clk = ~clk;

endmodule
```

Figure 10.3 Moore Detector Testbench

10.1.4 Behavioral Simulation

The *Moore1011.v* (Figure 10.2) design description file and its testbench, *Moore1011Tester.v* (Figure 10.3) are simulated in a ModelSim 5.7c project. The simulation run waveform is shown in Figure 10.4.

The present state of the machine (*p_state*) is displayed in the waveform shown. The output of the machine becomes **1** when *p_state* is **100** that correspond to state *e* of our Moore machine.

10.2 Design Implementation

This section discusses how Quartus II environment is used for hardware implementation of the Moore design of the previous section. Steps involved in this implementation include, project creation in Quartus II, generation of proper interfaces for our tested description of Moore machine, porting or adapting the Verilog code of our design to the Quartus II environment, design compilation, and device programming. After completion of these steps, using UP2 pushbuttons and LEDs, our UP2 prototype will be able to verify waveforms shown in Figure 10.4.

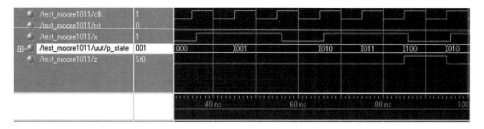

Figure 10.4 Moore Machine Simulation Run (Partial View)

10.2.1 Project Definition

The Quartus II project for hardware implementation of our Moore machine is *Detector*. This project uses our *BookLibrary* as its pre-defined user library. A user library can either be specified during creation of a project or in the *Settings* window after a project is created. To add a user library in the *Settings* window, open the *Assignments* pull-down menu from the Quartus II main window, and select *Settings*. When the *Settings* window opens, under *Files & Directories*, click on *User Libraries* to set your libraries.

10.2.2 Symbol Generation from Verilog Code

In order to be able to use our tested Verilog code of Figure 10.2 in Quartus, a symbol has to be created for it. The generated symbol becomes available in the *Project* library and can be used like any other symbol is a block diagram.

For generating a symbol for *Moore1011* module of Figure 10.2, copy its code to the directory of the Quartus II *Detector* project. Then open this file using Quartus II text editor (*File → Open*, and select *Moore1011.v* to open it). While this file is open, open the *File* pull-down menu, and follow links to create symbol for the current file. As with other symbols, we can edit the symbol that is so generated. The edited symbol for our *Moore1011* Verilog code is shown Figure 10.5.

10.2.3 Schematic Entry

The schematic file for Quartus II *Detector* project is *Detector.bdf*. In this schematic, the *moore1011* symbol of Figure 10.5, its necessary interfaces for UP2 resources, and the *Detector* pins are entered and proper interconnections made. Components from *BookLibrary* are used for interfacing pushbuttons of UP2 to the inputs of *moore1011* circuit. Figure 10.6 shows the complete schematic of the *Detector* circuit for programming a UP2 device.

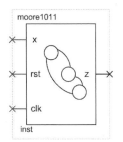

Figure 10.5 Moore Detector Symbol

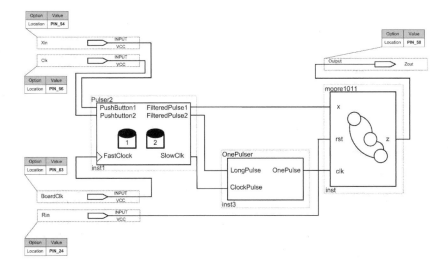

Figure 10.6 Detector System Description

After the completion of design entry phase, the design is compiled in Quartus II. This phase synthesizes the Verilog parts of the design and pulls together other cell-based and gate-level parts into a complete device programming file.

10.2.4 Compilation and Synthesis

By compiling the design of Figure 10.6, it is synthesized and proper programming files are created for it. In addition, various timing and floorplan views become available. For the *moore1011* component, a portion of its RTL view that is its post-synthesis netlist is shown in Figure 10.7. This netlist is generated for the FLEX 10K device being used as the synthesis target. This schematic shows the use of three flip-flops for implementing states of our detector circuit. AND-OR logic is used for flip-flop inputs.

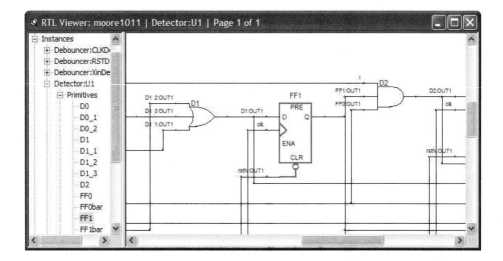

Figure 10.7 RTL View of *moore1011*

10.2.5 Device Programming and Testing

As previously discussed, either of the two UP2 devices can be programmed with the hardware of Figure 10.6. We have chosen the EPM7128S device for this design and Table 10.1 shows its corresponding pin settings. Pin numbers are also shown in Figure 10.6 in block IO tables. The complete design of *Detector*, including *moore1011* and its pushbutton interfaces, uses 34 of the 128 macrocells of the EPM7128S device.

For testing the design, the input and the clock are connected to UP2 pushbuttons and the circuit reset is connected to a UP2 switch. Use of a pushbutton for the clock allows manual slow operation of the circuit and thus, pushbuttons are debounced by using *Pulser2* from our *BookLibrary*. The debounced PB2 is put into *OnePulser* so that with each release of the pushbutton, a positive pulse is generated. This output provides the clock signal for *moore1011*. The other debounced output of *Pulser2* that corresponds to PB1 is directly connected to the Moore machine's *Xin* input (this is a level input). The circuit operates when its reset input switch is in the off position (SW1 must be down).

Table 10.1 *Detector* **Port Connections**

Detector Ports	MAX 7000S Pins	UP2 Resources
Xin	54	Wired to MAX PB 1
Rin	24	Wired to MAX SW 1
Clk	56	Wired to MAX PB2
BoardClock	83	25 MHz Clock
Zout	58	Wired to MAX LED 1

To test the circuit, hold down both pushbuttons and release and push PB2 several times to make sure the machine is in its initial state. Then release PB1 (this puts a **1** on *Xin*) and release and push PB2 to clock the circuit. Repeat this process for clocking a **0** and then two **1**s into the circuit through its *Xin* input. With the 4th release of PB2 (rising edge of the clock) the *a*-segment of SSD1 of MAX turn off, which means that *Zout* has become **1**.

Since the circuit is clocked and all *Xin* values are taken on the edge of *Clk*, we could test the circuit without using a debouncer for PB1 (the *Xin* input).

10.3 Summary

This chapter used a state machine example to demonstrate how a behavioral Verilog that is synthesizable could be used in a design and after synthesis incorporated with the other components of the design. We showed the procedure for simulating our behavioral design outside of Quartus II and after verifying it bringing it into Quartus II.

Part 3

System Design Projects

Having learned the use of Quartus II (and its related tools) for various forms of design entry, this part shows complete design projects. In these projects, hardware design aspects are emphasized and very little will be said about using specific design tools. Because of the complexity and the large size of these designs, we use the FLEX 10K device of UP2 for the projects of this part. Design projects presented here are:

- Sequential Multiplier
- VGA Adapter
- Keyboard Interface
- Design of Sayeh CPU

11 Sequential Multiplier

This chapter discusses the design, simulation and prototyping of a sequential multiplier. The multiplication process will be done by the shift-and-add sequential multiplication procedure. After a discussion of the multiplication method used, we present the details and interfacing of our design. Then the multiplier will be partitioned into its data and control parts, and each part will be designed separately. The completed design will be simulated in Verilog and tested by programming the FLEX 10K device of the UP2 board.

11.1 Sequential Multiplier Specification

The project is the design of a 2-bit sequential multiplier, with 8-bit A and B inputs and a 16-bit result. The block diagram of the circuit to be designed is shown in Figure 11.1. This multiplier has an 8-bit bi-directional I/O for inputting its A and B operands, and outputting its 16-bit output one byte at a time.

Multiplication begins with the *start* pulse, and the *databus* will contain operands A and B in two consecutive clock pulses. After accepting these data inputs, the multiplier begins its multiplication process and when it is completed, it starts sending the result out on the *databus*. When the least-significant byte is placed on *databus*, the *Lsb_out* output is issued, and for the most-significant byte, *msb_out* is issued. When both bytes are outputted, *done* becomes **1**, and the multiplier is ready for another set of data.

The multiplexed bi-directorial *databus* is used to reduce the total number of pins of the multiplier.

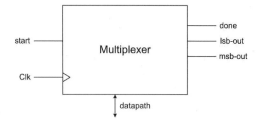

Figure 11.1 Multiplier Block Diagram

11.2 Shift-and-Add Multiplication

When designing multipliers there is always a compromise to be made between how fast the multiplication process is done and how much hardware we are using for its implementation.

A simple multiplication method that is slow, but efficient in use of hardware is the shift-and-add method. In this method, depending on bit i of operand A, either operand B is added to the collected partial result and then shifted to the right (when bit i is **1**), or (when bit i is **0**) the collected partial result is shifted one place to the right without being added to B.

This method is justified by considering how binary multiplication is done manually. Figure 11.2 shows manual multiplication of two 8-bit binary numbers.

We start considering bits of A from right to left. If a bit value is **0** we select 00000000 to be added with the next partial product, and if it is a **1**, the value of B is selected. This process repeats, but each time 00000000 or B is selected, it is written one place to the left with respect to the previous value. When all bits of A are considered, we add all calculated values to come up with the multiplication results.

Understanding hardware implementation of this procedure becomes easier if we make certain modifications to this procedure. First, instead of having to move our observation point from one bit of A to another, we put A in a shift-register, always observe its right-most bit, and after every calculation, we move it one place to the right, making its next bit accessible.

Second, for the partial products, instead of writing one and the next one to its left, when writing a partial product, we move it to the right as we are writing it, and the next one will not have to be shifted.

Finally, instead of calculating all partial products and at the end adding them up, when a partial product is calculated, we add it to the previous partial result and write the newly calculated value as the new partial result.

Therefore, if the bit of A that is being observed is **0**, 00000000 is to be added to the previously calculated partial result, and the new value should be shifted one place to the right. In this case, since the value being added to the partial result is *00000000*, adding is not necessary, and only shifting the partial result is sufficient. This process is called *shift*. However, if bit of A being observed is **1**, B is to be added to the previously calculated partial result, and the calculated new sum must be shifted one place to the right. This is called *add-and-shift.*

Repeating the above procedure, when all bits of A are shifted out, the partial result becomes the final multiplication result. We use a 4-bit example to clarify the above procedure. As shown in Figure 11.3, A = **1001** and B = **1101** are to be multiplied. Initially at time 0, A is in a shift-register with a register for partial results (P) on its left.

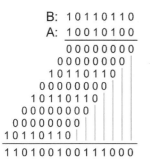

```
B:  1 0 1 1 0 1 1 0
A:  1 0 0 1 0 1 0 0
    0 0 0 0 0 0 0 0
    0 0 0 0 0 0 0 0
    1 0 1 1 0 1 1 0
    0 0 0 0 0 0 0 0
    1 0 1 1 0 1 1 0
    0 0 0 0 0 0 0 0
    0 0 0 0 0 0 0 0
    1 0 1 1 0 1 1 0
  1 1 0 1 0 0 1 0 0 1 1 1 0 0 0
```

Figure 11.2 Manual Binary Multiplication

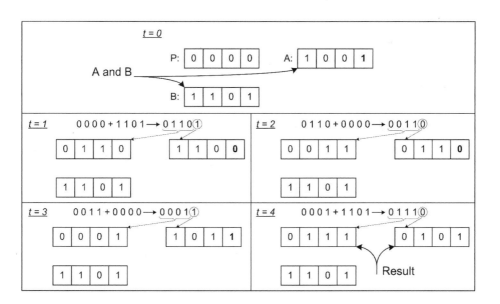

Figure 11.3 Hardware Oriented Multiplication Process

At the time 0, because *A[0]* is **1**, the partial sum of *B* + *P* is calculated. This value is **01101** (shown in the upper part of time 1) and has 5 bits to consider carry. The right most bit of this partial sum is shifted into the *A* register, and the other bits replace the old value of *P*. When *A* is shifted, **0** moves into the *A[0]* position. This value is observed at time 1. At this time, because *A[0]* is **0**, **0000** + *P* is calculated (instead of *B* + *P*). This value is **00110**, the right most bit of which is shifted into *A*, and the rest replace *P*. This process repeats 4 times, and at the end of the 4th cycle, the multiplication result becomes available in *P* and *A*. The least significant 4 bits of the result are in *A* and the most-significant bits are in *P*. The example used here performed 9*13 and 117 was obtained as the result of this operation.

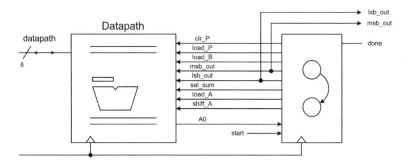

Figure 11.4 Data and Control Parts

11.3 Sequential Multiplier Design

The multiplication process discussed in the previous section justifies the hardware implementation that is being discussed here.

11.3.1 Control Data Partitioning

The multiplier has a datapath and a controller. The data part consists of registers, logic units and their interconnecting busses. The controller is a state machine that issues control signals for control of what gets clocked into the data registers.

As shown in Figure 11.4, the data path registers and the controller are triggered with the same clock signal. On the rising edge of a clock the controller goes into a new state. In this state, several control signals are issued, and as a result the components of the datapath start reacting to these signals. The time given for all activities of the datapath to stabilize is from one edge of the clock to another. Values that are propagated to the inputs of the datapath registers are clocked into these register with every clock edge.

11.3.2 Multiplier Datapath

Figure 11.5 shows the datapath of the sequential multiplier. As shown, P and B are 8-bit registers and A is an 8-bit shift-register. An adder, a multiplexer and a tri-state buffer constitute the other components of this datapath.

Control signals that are outputs of the controller and inputs of the datapath (Figure 11.4), are shown in bold in Figure 11.5 next to the data component that they control. These control signals control register clocking, bus assignments and logic unit output selections.

The input *databus* connects to the inputs of A and B to load multiplier and multiplicand into these registers. This bi-directional bus is driven by the output of P through an octal tri-state buffer, and by the tri-state output of A. This bi-directional bus is driven by the output of P through an actual tri-state buffer, and by the tri-state output of A. These tri-states become active when multiplication result is ready.

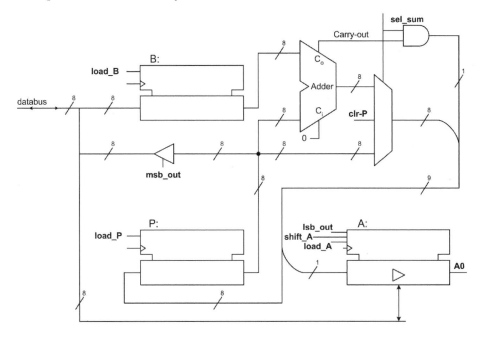

Figure 11.5 Multiplier Block Diagram

The output from B and P are put into an 8-bit adder for partial result in P to be added to B. The output of this adder $(P+B)$ feeds one side of a multiplexer. The other side of the multiplexer is driven by the P output, $(P+0)$. The *sel_sum* control input determines if $P+B$ or $P+0$ is to go on the multiplexer output.

The AND gate shown in Figure 11.5 selects carry-out from the adder or **0** depending of the value of *sel_sum* control input. This value is concatenated to the left of the multiplexer output to form a 9-bit vector. This vector has $P+B$ or $P+0$ with a carry to its left. The right-most bit of this 9-bit vector is split and goes into the serial input of the shift-register that contains A, and the other eight bits go into register P. Note that concatenation of the AND gate output to the left of multiplexer output and splitting the right bit from this 9-bit vector, effectively produces a shifted result that is clocked into P.

11.3.3 Description of Parts

Register P and B in Figure 11.5 are 8-bit registers with active high load-enable inputs. Module *Reg8*, shown in Figure 11.6 is used for these registers.

The adder used for adding P and B is a simple 8-bit adder with a carry-in and a carry-out and is shown in Figure 11.7. This description uses an **assign** statement that assigns $a+b+ci$ to the concatenation of co and s. With this assignment, the carry-out from the operation on the right-hand-side is captured in co.

Another component of the multiplier design is the 8-bit shift-register of Figure 11.8. The shift-register keeps its contents in its *im_data* intermediate variable. Depending on *{s1,s0}*, *im_data* is either untouched, shifted to the right, loaded with *data* or reset to 0.

```
module Reg8 ( d_in, clk, en, d_out );

    input [7:0] d_in;
    input clk, en;
    output [7:0] d_out;
    reg [7:0] d_out;

    always @( posedge clk )
        if (en) d_out = d_in;

endmodule
```

Figure 11.6 8-bit Register Used for *P* and *B*

```
module Add8 ( a, b, ci, s, co );

    input [7:0] a, b;
    input ci;
    output [7:0] s;
    output co;

    assign { co, s } = a + b + ci;

endmodule
```

Figure 11.7 8-bit Adder with Carry

```
module Shift8 ( clk, sin, s1, s0, oe, qa, data );

  input clk, sin, s1, s0, oe;
  output qa;
  inout [7:0] data;

  reg [7:0] im_data;

  always @( posedge clk )
    case ( { s1, s0 } )
      2'b00 : im_data = im_data;
      2'b01 : im_data = { sin, im_data[7:1] };
      2'b10 : im_data = data;
      2'b11 : im_data = 8'b00;
    endcase

  assign data = ( oe & ~s1 ) ? im_data : 8'hzz;
  assign qa = im_data[0];

endmodule
```

Figure 11.8 Shift-Register with Tri-state Output

```
module Mux8 ( a, b, sel, zero, y );

  input [7:0] a, b;
  input sel, zero;
  output [7:0] y;

  assign y = zero ? 8'h0 : ( ~sel ? a : b );

endmodule
```

Figure 11.9 Multiplexer

```
module Tri8 ( d_in, en, d_out );

  input [7:0] d_in;
  input en;
  output [7:0] d_out;

  assign d_out = en ? d_in : 8'hzz;

endmodule
```

Figure 11.10 Tri-state for Driving *databus*

When the output enable *(oe)* of the shift-register is active, *im_data* is placed on the *data* bi-directional port of the shift-register. Otherwise, *data* is float. Placement of *im_data* on *data* is also conditioned by *~s1*, so that *data* is driven only when not used as input.

Another component of the datapath of Figure 11.5 is the multiplexer of Figure 11.9. This multiplexer selects its *a* or *b* input depending on the value of *sel*. In addition, the multiplexer has a *zero* input that when **1**, it forces its output to *8'h0*. Since the multiplexer output connects to *P*, its zeroing feature is used for initial resetting of the *P* register.

As shown in Figure 11.5, an octal tri-state buffer connects the output of *P* to the bi-directional *databus*. The Verilog Code of this buffer is shown in Figure 11.10. The *en* input of this structure becomes active, when the most significant byte of the result that is in *P* is to go on the multiplier output *(databus)*.

11.3.4 Datapath Description

The Verilog Code of the *datapath* of the multiplier is shown in Figure 11.11. In this description components described above are instantiated and wired together according to the block diagram of Figure 11.5.

```
module datapath ( clk, clr_P, load_P, load_B, msb_out,
          lsb_out, sel_sum, load_A, shift_A, data, A0 );

    input clk, clr_P, load_P, load_B, msb_out, lsb_out, sel_sum, load_A, shift_A;
    inout [7:0] data;
    output A0;

    wire [7:0] B, P, sum, ShiftAdd;
    wire co;

    Reg8 latch_B ( data, clk, load_B, B );

    Add8 add_PB ( P, B, 1'b0, sum, co );

    Mux8 P_or_sum ( P, sum, sel_sum, clr_P, ShiftAdd );

    Reg8 latch_P ( {co&sel_sum,ShiftAdd[7:1]}, clk, load_P, P );

    Shift8 latch_A_shift ( clk, ShiftAdd[0], load_A, shift_A, lsb_out, A0, data );

    Tri8 buffer ( P, msb_out, data );

endmodule
```

Figure 11.11 Datapath Verilog Code

11.3.5 Multiplier Controller

The multiplier controller is a finite state machine that has two starting states, eight multiplication states, and two ending states. States and their binary assignments are shown in Figure 11.12. In the `idle` state the multiplier waits for `start` while loading A. In `init`, it loads the second operand B. In `m1` to `m8`, the multiplier performs add-and-shift of P+P, or P+0, depending on A0. In the last two states (`rslt1` and `1rslt2`), the two halves of the result are put on *databus*.

```
`define idle   4'b0000
`define init   4'b0001
`define m1     4'b0010
`define m2     4'b0011
`define m3     4'b0100
`define m4     4'b0101
`define m5     4'b0110
`define m6     4'b0111
`define m7     4'b1000
`define m8     4'b1001
`define rslt1  4'b1010
`define rslt2  4'b1011
```

Figure 11.12 Multiplier Control States

The Verilog Code of controller is shown in Figure 11.13. This Code declares *datapath* ports, and uses a single **always** block to issue control signals and make state transitions. At the beginning of this **always** block all control signal outputs are set to their inactive values. This eliminates unwanted latches that may be generated by the synthesis tool for these outputs.

The 4-bit *current* variable represents the currently active state of the machine. When *current* is `idle` and *start* is **0**, the *done* output remains high. In this state if *start* becomes **1**, control signals *load_A*, *clr_P* and *load_P* become active to load A with *databus* and clear the P register. Clearing P requires *clr_P* to put **0**'s on the multiplexer output by disabling it and loading the **0**'s into P by asserting *load_P*.

In `m1` to `m8` states, A is shifted, P is loaded, and if A0 is **1**, *sel_sum* is asserted. As discussed in relation to *datapath*, *sel_sum* controls shifted P+B or shifted P+0) to go into P.

In the result states *lsb_out* and *msb_out* are asserted in two consecutive clocks in order to put A and P on the *databus*, respectively.

```verilog
module controller ( clk, start, A0, clr_P, load_P, load_B,
          msb_out, lsb_out, sel_sum, load_A, Shift_A, done );

  input clk, start, A0;
  output clr_P, load_P, load_B, msb_out, lsb_out, sel_sum, done;
  output load_A, Shift_A;

  reg clr_P, load_P, load_B, msb_out, lsb_out, sel_sum, done;
  reg load_A, Shift_A;

  reg [3:0] current;

  always @ ( negedge clk ) begin
    clr_P = 0; load_P = 0; load_B = 0; msb_out = 0; lsb_out = 0; sel_sum = 0;
    load_A = 0; Shift_A = 0; done = 0;

    case ( current )
      `idle :
        if (~start) begin
          current = `idle;
          done = 1;
        end else begin
          current=`init;
          load_A = 1;
          clr_P = 1; load_P = 1;
        end
      `init :
        begin
          current=`m1;
          load_B = 1;
        end
      `m1, `m2, `m3, `m4, `m5, `m6, `m6, `m7, `m8 :
        begin
          current = current + 1; Shift_A = 1; load_P = 1;
          if (A0) sel_sum = 1;
        end
      `rslt1 :
        begin
          current=`rslt2; lsb_out = 1;
        end
      `rslt2 :
        begin
          current=`idle; msb_out = 1;
        end
      default : current=`idle;
    endcase

  end

endmodule
```

Figure 11.13 Verilog Code of Controller

```
module Multiplier ( clk, start, databus, lsb_out, msb_out, done );

   input clk, start;
   inout [7:0] databus;
   output done, lsb_out, msb_out;
   wire clr_P, load_P, load_B, msb_out, lsb_out, sel_sum, load_A, Shift_A;

   datapath dpu( clk, clr_P, load_P, load_B,
           msb_out, lsb_out, sel_sum, load_A, Shift_A, databus, A0 );
   controller cu( clk, start, A0, clr_P, load_P, load_B,
           msb_out, lsb_out, sel_sum, load_A, Shift_A, done );

endmodule
```

Figure 11.14 Top-Level Multiplier Code

11.3.6 Top-Level Code of the Multiplier

Figure 11.14 shows the top-level *Multiplier* **module**. The *datapath* and *controller* modules are instantiated here. The input and output ports of this unit are according to the diagram of Figure 11.1. This description is synthesizable, and can be ported into Quartus II for synthesis and device programming.

11.4 Multiplier Testing

This section shows an auto-check verifying testbench for our sequential multiplier. Several forms of data applications and result monitoring are demonstrated by this example. The outline of the *test_multiplier* **module** is shown in Figure 11.15.

In the declarative part of this testbench inputs of the multiplier are declared as **reg** and its outputs as **wire**. Since *databus* of the multiplier is a bidirectional bus, it is declared as **wire** for reading it, and a corresponding *im_data* **reg** is declared for writing into it. An **assign** statement drives *databus* with *im_data*. When writing into this bus from the testbench, the writing must be done into *im_data*, and after the completion of writing the bus must be released by writing 8'hzz into it.

Other variables declared in the testbench of Figure 11.15 are *expected_result* and *multiplier_result*. The latter is for the result read from the multiplier, and the former is what is calculated in the testbench. It is expected that these values are the same.

The testbench shown in Figure 11.15 applies three rounds of test to the *Multiplier* **module**. In each round, data is applied to the module under test and results are read and compared with the expected results. The following are tasks performed by this testbench:

- Read data files *data1.dat* and *data2.dat* and apply data to *databus*
- Apply *start* to start multiplication

- Calculate the expected result
- Wait for multiplication to complete, and collect the calculated result
- Compare expected and calculated results and issue error if they do not match

These tasks are independently timed, and at the same time, an **always** block generates a periodic signal on *clk* that clocks the multiplier.

```
`timescale 1ns/100ps

module test_multiplier;
  reg clk, start, error;
  wire [7:0] databus;
  wire lsb_out, msb_out, done;
  reg [7:0] mem1[0:2], mem2[0:2];
  reg [7:0] im_data, opnd1, opnd2;
  reg [15:0] expected_result, multiplier_result;
  integer indx;

  Multiplier uut ( clk, start, databus, lsb_out, msb_out, done );

  initial begin: Apply_data    . . .   end                    // Figure 11.16
  initial begin: Apply_Start    . . .   end                    // Figure 11.17
  initial begin: Expected_Result   . . .   end                 // Figure 11.18
  always @(posedge clk) begin: Actual_Result   . . .   end     // Figure 11.19
  always @(posedge clk) begin: Compare_Results   . . .   end // Figure 11.20
  always #50 clk = ~clk;
  assign databus=im_data;
endmodule
```

Figure 11.15 Multiplier Testbench Outline

11.4.1 Reading Data Files

Figure 11.16 shows the *Apply_data* **initial** block that is responsible for reading data and applying them to *im_data*, which in turn goes on *databus*. Data from *data1.dat* and *data2.dat* external lines are read into *mem1* and *mem2*. In each round of test data from *mem1* and *mem2* are put on *im_data*. Data from *mem2* is distanced from that of *mem1* by 100 ns. This way, the latter is interpreted as data for the *A* operand and the former for the *B* multiplication operand. After placing this data, 8'hzz is put on *im_data*. This releases the *databus* so that it can be driven by the multiplier when its result is ready.

11.4.2 Applying Start

Figure 11.17 shows an **initial** block in which variable initializations take place, and *start* signal is issued. Using a **repeat** statement, three 100 ns pulses distanced by 1350 ns are placed on *start*.

11.4.3 Calculating Expected Result

Figure 11.18 shows an **initial** block that reads data that is put on *databus* by the *Apply_data* block (Figure 11.16), and calculates the expected multiplication result. After *start*, when *databus* is updated, the first operand is read into *opnd1*. The next time *databus* changes, *opnd2* is read. The expected result is calculated using these operands.

```
initial begin: Apply_data
  indx=0;
  $readmemh( "data1.dat", mem1 );
  $readmemh( "data2.dat", mem2 );
  repeat(3) begin
    #300 im_data=mem1 [indx];
    #100 im_data=mem2 [indx];
    #100 im_data=8'hzz;
    indx=indx+1;
    #1000;
  end
  #200 $stop;
end
```

Figure 11.16 Reading Data Files

```
initial begin: Apply_Start
  clk=1'b0; start=1'b0; im_data=8'hzz;
  #200 ;
  repeat(3) begin
    #50   start=1'b1;
    #100  start=1'b0;
    #1350;
  end
end
```

Figure 11.17 Initializations and Start

```
initial begin: Expected_Result
  error=1'b0;
  repeat(3) begin
    wait ( start==1'b1 );
    @( databus );
    opnd1=databus;
    @( databus );
    opnd2=databus;
    expected_result = opnd1 * opnd2;
  end
end
```

Figure 11.18 Calculating Expected Result

11.4.4 Reading Multiplier Output

When the multiplier completes its task, it issues *msb_out* and *lsb_out* to signal that it has readied the two bytes of the result. The **always** block of Figure 11.19 is triggered by the rising edge of the circuit clock. After a clock edge, if *msb_out* or *lsb_out* is **1**, it reads the *databus* and puts in its corresponding position in *multiplier_result.*

```
always @(posedge clk) begin: Actual_Result
  if (msb_out) multiplier_result[15:8] = databus;
  if (lsb_out) multiplier_result[7:0] = databus;
end
```

Figure 11.19 Reading Multiplier Results

```
always @(posedge clk) begin: Compare_Results
  if (done)
    if (multiplier_result != expected_result) error = 1;
    else error = 0;
end
```

Figure 11.20 Comparing Results

11.4.5 Comparing Results

Figure 11.20 shows the **always** block that is responsible for comparing actual and expected multiplication results. After the active edge of the circuit clock if *done* is **1**, then comparing *multiplier_result* and *expected_result* takes place. If values of these variables do not match *error* is issued.

The self-running testbench presented here verifies RT-level operation of our multiplier. Prototyping this design using the UP2 board is presented in the next section.

11.5 Multiplier Prototyping

We use the FLEX 10K device of UP2 for prototyping our multiplier. This section describes porting the Verilog Code of the multiplier into Quartus II, generating switch and display interfaces for our design and programming the EPF10K70 of the UP2 development board.   The Quartus II project used for this part is *SeqMultiplier* in a design directory by the same name. *BookLibrary* is included in the list of libraries available to the project.

11.5.1 Porting Multiplier into Quartus II

The *Multiplier* **module** of Figure 11.14 is the top-level module of our multiplier. To be able to use this design in Quartus II, this and all its related Verilog Files must be copied to the directory of the *SeqMultiplier* project.

In order to use the *Multiplier* **module** in a Quartus II schematic, a symbol has to be created for it. Figure 11.21 shows this symbol created by Quartus II, after some manual editings.

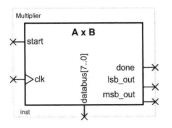

Figure 11.21 Multiplier Symbol

11.5.2 Multiplier Interfaces

Figure 11.22 shows the *SeqMultiplier* schematic that includes the *Multiplier* and its pushbutton and display interfaces. In order to step through the multiplication process, its clock is driven by a pushbutton. The other pushbutton available to FLEX is used for the *start* input. FLEX switch set is used for the *A* and *B* operands. We manually set these switches to values that are to be multiplied.

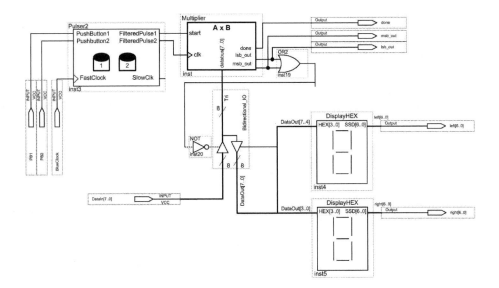

Figure 11.22 *SeqMultiplier* Prototype

For the display of the output of the multiplier two instances of *DisplayHEX* from *BookLibrary* are used. These outputs display both halves of the 16-bit output of the multiplier.

11.5.3 Bidirectional Databus

The multiplier *databus* is a bidirectional bus used for *A* and *B* operands as well as the two halves of the result. We have used the *lpm_bustri* megafunction of Quartus II, that is available under *gates* category of megafunctions, to split the in-side and out-side of the *databus*.

The FLEX switches connect to the input side of *Bidirectional-IO* component, and the displays connect to its output side. When either *lsb_out* or *msb_out* is issued by the multiplier, the *databus* connects to the displays through the *Bidirectional-IO*. At all other times that the multiplier is not driving its output, the switches drive the *databus*.

11.5.4 Operating the Prototype

Compiling *SeqMultiplier* of Figure 11.22 synthesizes the *Multiplier* **module** and together with the rest of components of this design, generates the *SeqMultiplier.sof* file for programming the FLEX 10K device.

Pin assignments are done according to permanently assigned pins of FLEX. Bits of *DataIn* port of diagram of Figure 11.22 are connected to the switches according to Figure 6.34. The outputs of the *DisplayHEX* components are assigned to the seven segment displays according to Figure 6.35, and inputs PB1 and PB2 are assigned to FLEX pushbuttons as shown in Figure 6.33.

To test the multiplier, the switches are set to a test value for *A* and while *start* is **1**, a clock pulse is given. Then, while *start* is **0**, the switches are set to a value for *B* and another clock pulse is given. Following the leading of *A* and *B*, eight clock pulses are given (releasing and pressing PB2 eight times) to complete the multiplication process. With the next clock, the right-most byte of the result becomes available on the SSDs, and with the next clock the left-most byte of the multiplication result becomes available on the SSDs. Both values of the output are displayed in hexadecimal Code.

Figure 11.23 shows part of the FLEX 10K timing closure floorplan after being programmed with our multiplier. The complete *SeqMultiplier* project uses 230 Logic Elements of the total 3744 available on FLEX 10K. Of the available 36,804 memory bits, none are used. The timing viewer allows cells to be selected and timing between them be viewed. When a cell is selected, its fan-ins and fan-outs are listed and corresponding delay values are shown on the arrows going between the logic elements; Figure 11.23 shows an example.

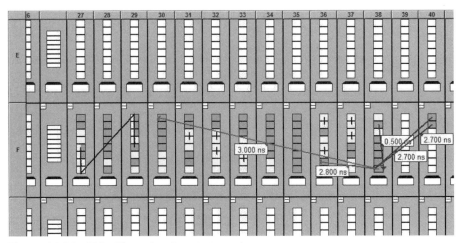

Figure 11.23 Chip Floorplan (Partial View)

11.6 Summary

This chapter showed a complete design of a system with a well-defined datapath and a good-size controller. The design demonstrates top-down design and data/control partitioning. We showed how this design could be implemented by coding lower level RTL parts and then wiring them into a complete system. Concepts of controllers, control signals controlling data activities, bussing, and various forms of unidirectional and bi-directional busses were demonstrated in this design. We demonstrated how the UP2 board could be utilized to test the physical implementation of an HDL based design.

12 VGA Adapter

The design of a VGA adapter capable of displaying characters from a display RAM on a standard VGA monitor is discussed in this chapter. The chapter discusses the basics of interfacing with a VGA monitor and develops an interface using the UP2 development board. Unlike the previous chapter that dealt with a top-down data/control design, this chapter designs small interfaces and puts a complete design together. The design methodology presented here uses Verilog blocks, megafunctions, memories, and schematic capture to complete the design of a display adapter.

12.1 VGA Driver Operation

A standard VGA monitor consists of a grid of pixels that can be divided into rows and columns. A VGA monitor contains at least 480 rows, with 640 pixels per row, as shown in Figure 12.1. Each pixel can display various colors, depending on the state of the red, green, and blue signals.

Each VGA monitor has an internal clock that determines when each pixel is updated. This clock operates at the VGA-specified frequency of 25.175 MHz. The monitor refreshes the screen in a prescribed manner that is partially controlled by the horizontal and vertical synchronization signals. The monitor starts each refresh cycle by updating the pixel in the top left-hand corner of the screen, which can be treated as the origin of an X–Y plane (see Figure 12.1).

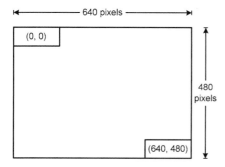

Figure 12.1 VGA Monitor

After the first pixel is refreshed, the monitor refreshes the remaining pixels in the row. When the monitor receives a pulse on the horizontal synchronization, it refreshes the next row of pixels. This process is repeated until the monitor reaches the bottom of the screen. When the monitor reaches the bottom of the screen, the vertical synchronization pulses cause the monitor to begin refreshing pixels at the top of the screen (i.e., at [0,0]).

12.1.1 VGA Timing

For the VGA monitor to work properly, it must receive data at specific times with specific pulses. Horizontal and vertical synchronization pulses must occur at specified times to synchronize the monitor while it is receiving color data.

Figure 12.2 shows the timing waveform for the color information with respect to the horizontal synchronization signal. Based on the clock frequency, these times translate to certain number of clock cycles shown in this figure. For example, a horizontal sweep (parameter A) that takes 31.77 µs, translates to 800 clock cycles of 25.175 MHz.

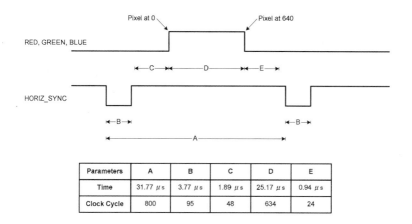

Figure 12.2 Horizontal Refresh Cycle

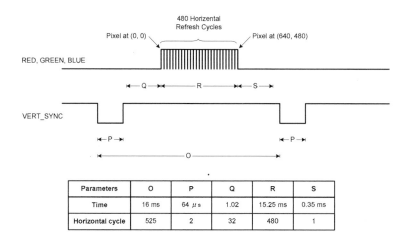

Parameters	O	P	Q	R	S
Time	16 ms	64 μs	1.02	15.25 ms	0.35 ms
Horizontal cycle	525	2	32	480	1

Figure 12.3 Vertical Refresh Cycle

Figure 12.3 shows the timing waveform for the color information with respect to the vertical synchronization signal. Based on the fact that a horizontal sweep takes 31.77 µs (800 clock cycles), the times shown take a certain number of horizontal refresh cycles that are shown in this figure. For example, a screen refresh cycle that takes 16.7 ms, translates to 525 horizontal cycles of 31.77 µs.

The frequency of operation and the number of pixels that the monitor must update determine the time required for updating each pixel, and the time required for updating the whole screen. The following equations roughly calculate the time required for the monitor to perform all of its functions.

$$T_{pixel} = 1/f_{CLK} = 40 \text{ ns}$$
$$T_{ROW} = A = B + C + D + E$$
$$= (T_{pixel} \cdot 640 \text{ pixels}) + \text{row} + \text{guard bands} = 31.77 \text{ µs}$$
$$T_{screen} = O = P + Q + R + S$$
$$= (T_{ROW} \cdot 480 \text{ rows}) + \text{guard bands} = 16.6 \text{ ms}$$

Where:

T_{pixel}=Time required to update a pixel
f_{CLK}=25.175 MHz
T_{ROW}=Time required to update one row
T_{screen}=Time required to update the screen
B, C, E=Guard bands
P, Q, S=Guard bands

The monitor writes to the screen by sending red, green, blue, horizontal and vertical synchronization signals when the screen is at the expected location.

Once the timing of the horizontal and vertical synchronization signals is accurate, the monitor only needs to keep track of the current location, so it can send the correct color data to the pixel.

12.1.2 Monitor Synchronization Hardware

The hardware required for VGA signal generation must keep track of the number of 25.175 MHz clock cycles, and issue signals according to the timing waveforms of Figure 12.2 and Figure 12.3. The Verilog code of Figure 12.4 uses the *SynchClock* clock signal to generate *Hsynch* (HORIZ_SYNCH of Figure 12.2) and *Vsynch* (VERT_SYNCH of Figure 12.3).

```verilog
module MonitorSynch
(     RedIn, GreenIn, BlueIn, SynchClock, Red, Green, Blue, Hsynch, PixelRow, PixelCol, Vsynch
);
      input RedIn, GreenIn, BlueIn, SynchClock;
      output Red, Green, Blue, Hsynch, Vsynch;
      output [9:0] PixelRow, PixelCol;

      reg Red, Green, Blue, Vsynch, Hsynch;
      reg [9:0] PixelRow, PixelCol;
      reg [9:0] Hcount, Vcount;

      always @(posedge SynchClock) begin
          if (Hcount == 799) Hcount =0;
              else Hcount = Hcount + 1;
          if (Hcount >= 661 && Hcount <= 756) Hsynch = 0;
              else Hsynch = 1;
          if (Vcount >= 525 && Hcount >= 756) Vcount = 0;
              else if (Hcount == 756) Vcount = Vcount + 1;
          if (Vcount >= 491 && Vcount <= 493) Vsynch = 0;
              else Vsynch = 1;

          if (Hcount <= 640) PixelCol = Hcount;
          if (Vcount <= 480) PixelRow = Vcount;

          if (Hcount <= 680 && Vcount <= 480) begin
              Red = RedIn;  Green = GreenIn;  Blue = BlueIn;
          end else
              {Red, Green, Blue} = 0;
      end
endmodule
```

Figure 12.4 Monitor Synchronization Verilog Code

The code shown, uses color specifications from *RedIn, GreenIn* and *BlueIn* input signals and during the time periods specified by parameter *D* in Figure 12.2 and parameter *R* in Figure 12.3, puts them on the *Red, Green* and *Blue* output signals. At any point in time, the Verilog code of Figure 12.4 outputs the position of the pixel being updated in its 10-bit *PixelRow* and *PixelCol* output vectors.

The *Hcount* variable in this Verilog code keeps track of the number of clock cycles in each row, and *Vcount* is the number of horizontal cycles in each screen. Considering the very first pixel at (0, 0) position, the counting of the horizontal pixels begins at the beginning of the *D* region of the waveform of Figure 12.2. Therefore as the Verilog code shows, *Hsynch* becomes **0** when *Hcount* is between 661 and 756 (this is the *B* region). Likewise, considering the beginning of region *R* as the 0 point, the *P* region in Figure 12.3 begins at *Vcount* of 491 and ends at 493. Therefore, as the code shows, *Vsynch* is **0** during such *Vcount* values. With the (0, 0) point defined as such, pixels are active while *Hcount* is between 0 and 640 and *Vcount* is between 0 and 480. During these count periods output colors are active and *PixelRow* and *PixelCol* outputs reflect *Hcount* and *Vcount* respectively.

The Verilog code of Figure 12.4 is defined as a block shown in Figure 12.5 to be used in our implementation of a character display design.

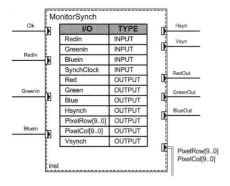

Figure 12.5 Monitor Synchronization Block Specification

12.2 Character Display

The design we are considering in this section is a character display hardware that outputs an address to a character display memory and inputs an ASCII code representing the character to display. We assume the display memory has 300 ASCII characters that will be displayed in 15 rows of 20 characters. Considering the 480 by 640 resolution, this makes each character occupy a matrix of 32×32 pixels.

In addition to the synchronization module of the previous section, the character display hardware has a character matrix and a pixel generation module. The character matrix defines active pixels of the supported characters and the pixel generation module reads this matrix and produces active color inputs (*RedIn*, *GreenIn*, and *BlueIn*) for the synchronization module.

12.2.1 Character Matrix

In our simple design we use 8×8 character resolution and only support ASCII characters from 32 to 95. With these 64 supported characters, our character matrix becomes an 8-bit memory of 512 words, in which every 8 consecutive words define a character. For example, as shown in Figure 12.6, pixels for character "5" with ASCII code of 53 decimal, begin at address 0A8 Hex that is (53-32)×8.

```
0A8  :  01111110 ; %      ******      %
0A9  :  01100000 ; %      **          %
0AA  :  01111100 ; %      *****       %
0AB  :  00000110 ; %            **    %
0AC  :  00000110 ; %            **    %
0AD  :  01100110 ; %      **    **    %
0AE  :  00111100 ; %       ****       %
0AF  :  00000000 ; %                  %
```

Figure 12.6 Character Matrix for Character "5"

```
DEPTH = 512;
WIDTH = 8;
ADDRESS_RADIX = HEX;
DATA_RADIX = BIN;
% Character Matrix ROM, addressed by PixelGeneration module %
CONTENT

BEGIN

% ASCII 0010_0000 to 0010_1111   %
000  :  00000000 ; %             %
001  :  00000000 ; %             %
002  :  00000000 ; %             %
003  :  00000000 ; %             %
004  :  00000000 ; %             %
005  :  00000000 ; %             %
006  :  00000000 ; %             %
007  :  00000000 ; %             %
 .  .  .
1F8  :  00000000 ; %             %
1F9  :  00010000 ; %        *    %
1FA  :  00110000 ; %       **    %
1FB  :  01111111 ; %     *******  %
1FC  :  01111111 ; %     *******  %
1FD  :  00110000 ; %       **    %
1FE  :  00010000 ; %        *    %
1FF  :  00000000 ; %             %

END;
```

Figure 12.7 Character Matrix *mif* File

For implementing the character matrix we use the LPM_ROM megafunction of Quartus II. This component is available under the *storage* category of megafunctions. With the aid of the megafunction wizard, this component is configured as an 8-bit memory with 9 address lines. During the configuration process we are asked to enter the Memory Initialization File name (*.mif*), for which we use *CharMtx.mif*. Using the *mif* format, pixel values (similar to those shown in Figure 12.6 for character "5") are defined for ASCII characters from 32 to 95. Figure 12.7 shows the beginning and end of this file, from which the formatting can be seen.

The symbol of the *CharMtx* component defined in Quartus II is shown in Figure 12.8. The input of this ROM is a 9-bit address and its output is the horizontal slice of the display code of the character. The most significant 6 bits of the address are the ASCII code for the display character, and bits 2 to 0 of the address determine its slice number.

Figure 12.8 Character Matrix Symbol

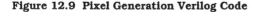

```verilog
module PixelGeneration (PixelRow, PixelCol, Clk, Char, MtxPntr, CharPntr);
    input [9:0] PixelRow, PixelCol;
    input Clk;
    input [7:0] Char;
    output [8:0] MtxPntr,
    output [8:0] CharPntr;

    reg [5:0] MtxStart;
    reg [8:0] MtxPntr;
    reg [8:0] CharPntr;

    wire [4:0] ScreenLine, ScreenPos;
    assign ScreenLine = PixelRow [9:5]; // 15 Lines=480/32
    assign ScreenPos = PixelCol [9:5];  // 20 Positions=640/32

    always @(posedge Clk) begin
        CharPntr = ScreenLine*20 + ScreenPos;
        MtxStart = Char - 32;
        // Char resolution is 8 pixel rows, bits [4:2];
        MtxPntr = {MtxStart, PixelRow[4:2]};
    end
endmodule
```

Figure 12.9 Pixel Generation Verilog Code

12.2.2   Pixel Generation Module

Another component for our character display hardware is the *PixelGeneration* module. This module uses the 640×480 pixel coordinates from the *MonitorSynch* module and translates it to our low-resolution 20×15 character coordinates. Using the latter coordinates it generates an address for our display memory (that contains 300 ASCII codes) and reads the corresponding ASCII code. This code, offset by 32, and the pixel position that is being refreshed determine a pointer for the character matrix (*CharMtx*) discussed above. The Verilog code of this module is shown in Figure 12.9.

The output of the *CharMtx* ROM is an 8-bit slice of the character being displayed. The specific bit that is to be displayed is selected by pixel column position coming from the *MonitorSynch* module.

12.2.3   Character Display Hardware

The complete schematic diagram of the character display hardware is shown in Figure 12.10. This hardware is our VGA adapter that addresses a memory of 300 characters, reads the character and displays it in one of the 300 locations of the screen. This hardware uses *MonitorSynch*, *PixelGeneration*, *CharMtx*, and an 8-to-1 multiplexer.

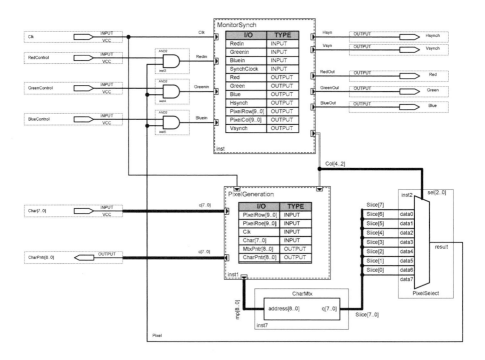

Figure 12.10   VGA Adapter Complete Schematic

The *MonitorSynch* module continuously sweeps across the 640×480 pixel screen and refreshes pixels with colors specified by its three color inputs. At the same time it reports the position of the pixel being refreshed to *PixelGeneration*. Based on these coordinate, this module calculates the address of the character that is being displayed, and using the *CharMtx* and the 8-to-1 multiplexer determines the value of the pixel in the screen position being refreshed. This pixel value allows color inputs to be used by the *MonitorSynch* module for painting the pixel.

12.3 UP2 Prototyping

The design of our VGA adapter of Figure 12.10 is complete in the sense that it addresses a character and displays it on the monitor. Testing this design requires a display memory with some data. Figure 12.11 shows a system utilizing this *Adapter* circuit to display character contents of *DisplayMem* while providing a mechanism of writing new characters into this memory.

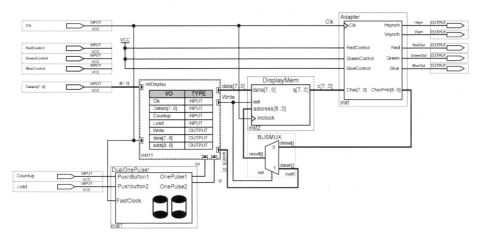

Figure 12.11 Prototyping the VGA Adapter (*Adapter*)

12.3.1 Display Memory

The *DisplayMem* block is a memory block of 512 8-bit words, of which only 300 are used. This memory is addressed by *Adapter* for reading from it, and by *InitDisplay* for writing into it. When writing into the memory, the *BUSMUX* multiplexer selects *addr* address output of *InitDisplay*. Writing into this memory is clocked, for which the main system clock is used, while reading is done asynchronous.

The display memory is designed by configuring the *LPM_RAM_DQ* megafunction of Quartus II. As with other memories, this megafunction is in the *storage* category of megafunctions. While configuring it, a memory initialization file in the *mif* format is specified. This file contains test data that will be displayed. Also, during configuration of *DisplayMem*, input clocking, write-enable and other memory parameters will be specified.

12.3.2 Address Selection

The *BUSMUX* multiplexer that selects the read or write address of *DiplayMem* is a 9-bit 2-to-1 multiplexer. This unit is a megafunction in Quartus II.

12.3.3 Writing Display Data

The *InitDisplay* Verilog module of Figure 12.11 has a counter that counts between 0 and 299 with its *Countup* input and loads its *DataIn* input in its *data* output when *Load* is issued. We use this circuit to count to a location in *DisplayMem* and load data from UP2 switches into this memory. If *Load* is pressed twice in a row, the counter resets to location 0. Count-up and load input are taken from the UP2 pushbuttons.

```
module InitDisplay ( Clk, DataIn, Countup, Load, Write, data, addr );
    input Clk;
    input [7:0] DataIn;
    input Countup, Load;
    output Write;
    output [7:0] data;
    output [8:0] addr;

    reg Write;
    reg [8:0] addr;
    reg loaded; //two Loads in a row resets

    assign data = DataIn;

    always @(posedge Clk)
        if (Countup) begin
            addr = addr + 1;
            if (addr == 300) addr = 0;
            Write = 0;
            loaded = 0;
        end else if (Load) begin
            if (loaded) begin // perform reset
                addr = 0;
                loaded = 0;
            end else begin  // perform loading
                Write = 1;
                loaded = 1;
            end
        end else Write = 0;
endmodule
```

Figure 12.12 Module for Manual Loading of Display Memory

InitDisplay is implemented in Verilog in a schematic block of Quartus II. As shown in the Verilog code of Figure 12.12, *DataIn* continuously drives *data* that is the data to be written into the display memory. When *Countup* is issued, *addr* is incremented. When *Load* is issued, the output *Write* (write-enable) becomes active, that causes a write at the *addr* location.

12.3.4 Pushbutton Interfaces

The *DualOnePulsers* component of Figure 12.12 uses a dual debouncer (Figure 8.14) and two one-pulse generators (Figure 8.12) to generate synchronous one-clock duration pulses for every time a pushbutton is pressed.

12.3.5 Pin Assignments

In order to test out design, data and control inputs of Figure 12.11 are connected to UP2 switches and pushbuttons. The outputs of this circuit should be connected to FLEX 10K pins according to connections shown in Figure 6.36. Figure 12.13 shows these pin assignments.

12.3.6 Prototype Operation

Upon programming the FLEX 10K device, the monitor connected to VGA D-sub connector displays characters in the *DisplayMem* file. We can write any ASCII character on the display by setting its ASCII code on the FLEX switches, pressing the *Countup* pushbutton some number of times and then pressing *Load* to write the ASCII character in the counted location of the screen.

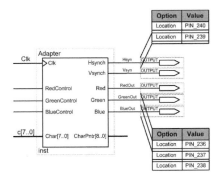

Figure 12.13 Adapter Pin Assignments to the VGA Connector

Note that our hardware only supports ASCII characters from 32 to 95, and large characters are displayed. To start from location 0, press *Load* twice.

12.4 Summary

This chapter showed a complete design by use of Verilog, schematics and megafunctions. Some features of Quartus II, such as use of memory blocks that were not discussed before, were presented in this chapter. If done properly, the use of memory blocks is an Altera design uses FPGA memory bits that can free up a large number of logic elements for other uses. Dual-port memories cannot be implemented with FLEX memory bits and must be implemented using logic element flip-flops. Not only this is an inefficient use of logic elements, such memories cannot be initialized with memory initialization files.

In addition to presenting an elaborate use of Quartus II, this chapter showed the design of a VGA adapter. Understanding display monitors and being able to program them is important for logic designers and students in the digital field.

13 Keyboard Interface

This chapter deals with the design of a keyboard interface that is implemented on UP2 using its FLEX 10K device. The chapter discusses how keyboards work and how they transmit data to and receive data from a computer. We will then take a simplified approach and show the interface for receiving data from a keyboard. The interface receives serial data from the keyboard and generates ASCII codes for keys that ASCII codes are applicable.

13.1 Data Transmission

Data communication between the keyboard and the host system is synchronous serial over bi-directional clock and data lines. Keyboard sends commands and key codes, and the system sends commands to the keyboard.

Either the system or the keyboard drive the data and clock lines, while clocking data in either direction is provided by the keyboard clock. When no communication is occurring, both lines are high. Figure 13.1 shows the timing of keyboard serial data transmission.

13.1.1 Serial Data Format

Data transmission on the data line is synchronized with the clock; data will be valid before the falling edge and after the rising edge of the clock pulse. Serial data transmission begins with the data line dropping to 0. This bit value is taken on the rising edge of the clock and considered as the start-bit. On the next eight clock edges, data is transmitted in low to high order bit. The next data bit is the odd-parity bit, such that data bits and the parity bit always have odd number of ones. The last bit on the data line is the stop-bit that is always

1. After the stop-bit, the data line remains high until another transmission begins.

When the keyboard sends data to or receives data from the system it generates the clock signal to time the data. The system can prevent the keyboard from sending data by forcing the clock line to **0**, during this time the data line may be high or low. When the system sends data to the keyboard, it forces the data line to **0** until the keyboard starts to clock the data stream.

13.1.2 Keyboard Transmission

When the keyboard is ready to send data, it first checks the status of the clock to see if it is allowed to transmit data. If the clock line is forced to low by the system, data transmission to the system is inhibited and keyboard data is stored in the keyboard buffer. If the clock line is high and the data line is low, the keyboard is to receive data from the system. In this case, keyboard data is stored in the keyboard buffer, and the keyboard receives system data. If the clock and data lines are both high the keyboard sends the start-bit, 8 data bits, the parity bit and the stop-bit.

During transmission, the keyboard checks the clock line for low level at least every 60 μseconds. If the system forces the clock line to **0** after the keyboard starts ending data, a condition known as line contention occurs, and the keyboard stops sending data. If line contention occurs before the rising edge of the 10th clock pulse, the keyboard buffer returns the clock and data lines to high level.

13.1.3 System Transmission

The system sends 8-bit commands to the keyboard. When the system is ready to send a command to the keyboard, it first checks to see if the keyboard is sending data. If the keyboard is sending, but has not reached the 10th clock signal, the system can override the keyboard output by forcing the keyboard clock line to **0**. If the keyboard transmission is beyond the 10th clock signal, the system receives the transmission.

If the keyboard is not sending or if the system decides to override the output of the keyboard, the system forces the keyboard clock line to **0** for more than 60 μseconds while preparing to send data. When the system is ready to send the start bit, it allows the keyboard to drive the clock line to **1** and drives the data line to low. This signals the keyboard that data is being transmitted from the system. The keyboard generates the clock signals and receives the data bits, parity and the stop-bit. After the stop-bit, the system releases the data line. If the keyboard receives the stop-bit it forces the data line low to signal the system that the keyboard has received its data.

Upon receipt of this signal, the system returns to a ready state, in which it can accept keyboard output or goes to the inhibited state until it is ready. If the keyboard does not receive the stop-bit, a framing error has occurred, and the keyboard continues to generate clock signals until the data line becomes high. The keyboard then makes the data line low and requests a resending of the data. A parity error will also generate a re-send request by the keyboard.

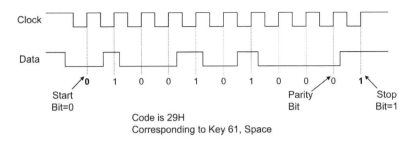

Figure 13.1 Keyboard Serial Data Transmission

13.1.4 Power-On Routine

The keyboard logic generates a power-on-reset signal when power is first applied to the keyboard. The timing of this signaling is between 150 milliseconds and 2.0 seconds from the time power is first applied to the keyboard.

Following this signaling, basic assurance test is performed by the keyboard. This test consists of a keyboard processor test, a checksum of its ROM, and a RAM test. During this test, activities on the clock and data lines are ignored. The keyboard LEDs are turned on at the beginning and off at the end of the test. This test takes a minimum of 300 milliseconds and a maximum of 500 milliseconds. Upon satisfactory completion of the basic assurance test, a completion code (hex AA) is sent to the system, and keyboard scanning begins.

Table 13.1 System Commands to Keyboard

Command	Hex
Set/Reset Status Indicators	ED
Echo	EE
Invalid Command	EF
Select Alternate Scan Codes	F0
Invalid Command	F1
Read ID	F2
Set Typematic Rate/Delay	F3
Enable	F4
Default Disable	F5
Set Default	F6
Set All Keys – Typematic	F7
- Make/Break	F8
- Make	F9
- Typematic/Make/Break	FA
Set Key Type – Typematic	FB
- Make/Break	FC
- Make	FD
Resend	FE
Reset	FF

13.2 Codes and Commands

A host system may send 8-bit commands to the keyboard, while a keyboard may send commands and key codes to the system.

13.2.1 System Commands

System commands may be sent to the keyboard at any time. The keyboard will respond within 20 milliseconds, except when performing the basic assurance test (BAT), or executing a Reset command.   System commands and their hexadecimal values are shown in Table 13.1.

13.2.2 Keyboard Commands

Table 13.2 shows the commands that the keyboard may send to the system and their hexadecimal values.

Table 13.2 Keyboard Commands to System

Command	Hex
Key Detection Error/Overrun	00
Keyboard ID	83AB
BAT Completion Code	AA
BAT Failure Code	FC
Echo	EE
Acknowledge (ACK)	FA
Resent	FE

13.2.3 Keyboard Codes

Keyboards are available for several languages and settings. The keyboard that is most common for the English language is one with 104 keys shown in Figure 13.2. Keys of this keyboard are identified by numbers, and for every key there is a scan code. Several scan codes are available, and the default scan code is Scan Code 2 that we will discuss here.

Keyboard scan codes consist of a Make and a Break code. The Make code identifies the key pressed and the Break code indicates the release of a key. For most keys the Break code is F0 followed by the Make code. For example when the Space bar (key 61) is pressed and released, hexadecimal codes 29, F0 and 29 are transmitted from the keyboard to the system via the data serial line. If this key remains pressed, the Make code (29) is continuously transmitted until it is released. Make codes for Scan Code 2 are shown in Table 13.3

The Make and Break arrangement, makes it possible for the system to identify multiple keys pressed and the order in which they have been pressed. For example, if one presses and holds down the Left-Shift key (key number 44), 12 Hex is continuously sent to the system. While this is happening, if key number 9 (the 8/* key) is pressed and released, 3E, F0 and 3E codes are transmitted.   The receiving system identifies this sequence of events as the intention to enter an asterisk (*).

Figure 13.2 Standard 104-key Keyboard and Key Numbers

Table 13.3 Keyboard Scan Codes and Corresponding ASCII Characters

Key Numb	Make Code	ASCII No Shift	ASCII Shift	Character No Shift	Character Shift	Key Numb	Make Code	ASCII No Shift	ASCII Shift	Character No Shift	Character Shift
1	0E	96	126	`	~	34	2B	70	102	F	f
2	16	49	33	1	!	35	34	71	103	G	g
3	1E	50	64	2	@	36	33	72	104	H	h
4	26	51	35	3	#	37	3B	74	106	J	j
5	25	52	36	4	$	38	42	75	107	K	k
6	2E	53	37	5	%	39	4B	76	108	L	l
7	36	54	94	6	^	40	4C	59	58	;	:
8	3D	55	38	7	&	41	52	39	34	'	"
9	3E	56	42	8	*	42	5D				
10	46	57	40	9	(	43	5A	13	Enter	Enter	Enter
11	45	48	41	0	)	44	12		Shift	Shift	Shift
12	4E	45	95	-	_	45	61				
13	55	61	43	=	+	46	1A	90	122	Z	z
15	66	08	08	BS	BS	47	22	88	120	X	x
16	0D	09	09	Tab	Tab	48	21	67	99	C	c
17	15	81	113	Q	q	49	2A	86	118	V	v
18	1D	87	119	W	w	50	32	66	98	B	b
19	24	69	101	E	e	51	31	78	110	N	n
20	2D	82	114	R	r	52	3A	77	109	M	m
21	2C	84	116	T	t	53	41	44	60	,	<
22	35	89	121	Y	y	54	49	46	62	.	>
23	3C	85	117	U	u	55	4A	47	63	/	?
24	43	73	105	I	i	57	59			Shift	Shift
25	44	79	111	O	o	58	14			Ctrl	Ctrl
26	4D	80	112	P	p	59	E0 1F			Win	Win
27	54	91	123	[	{	60	11			Alt	Alt
28	5B	93	125	]	}	61	29	32	32	Space	Space
29	5D	92	124	\			62	E0/11			Alt	Alt
30	58			Caps	Caps	63	E0 27			Win	Win
31	1C	65	97	A	a	64	E0/14			Ctrl	Ctrl
32	1B	83	115	S	s	65	E0 2F			Menu	Menu
33	23	68	100	D	d						

13.3 Keyboard Interface Design

This section discusses a keyboard interface for reading scan data from the keyboard and producing ASCII codes of the keys pressed. Code Set 2 is assumed, and the interface only handles data transmission from the keyboard. The interface reads serial data from the keyboard, detects the Make code when a key is pressed and looks up the Make code in an ASCII conversion table. For simplicity, the look-up table only handles upper-case characters.

13.3.1 Collecting the Make Code

The first part of our interface connects to the keyboard data and clock lines and when a key is pressed, it outputs an 8-bit scan code. The *KBdata, KBclock* inputs are for the keyboard data and clock inputs, and the 8-bit *ScanCode* is the main output of this part.

This part also uses a fast synchronizing clock, *SYNclk*, and a keyboard reset input, *KBreset*. In addition to the *ScanCode* output, this part outputs a signal to indicate that a scan code is ready (*ScanRdy*) and another output to indicate that a key has been released (*KeyReleased*). These outputs make distinction between Make and Break states.

```
module KeyboardInterface
  (KBclk, KBdata, ResetKB, SYNclk, ScanRdy, ScanCode, KeyReleased);
  input KBclk;
  input KBdata;
  input ResetKB;
  input ReadKB;
  input SYNclk;
  output ScanRdy;
  output [7:0] ScanCode;
  output KeyReleased;

  // Details in Figure 13.4
  // Generate an internal synchronized clock
  always @(posedge Clock) begin
    . . .
    // Count the number of serial bits and collect data into ScanCode
    . . .
  end

  // Details in Figure 13.5
  always @(posedge SYNclk) begin
    . . .
    // Keep track of the state of Scan Codes outputted
    . . .
  end
  // Issue KeyReleased when done

endmodule
```

Figure 13.3 Verilog Pseudo Code

The pseudo-code of this unit is shown in Figure 13.3.   After the declarations, in this part an internal clock (*Clock*) that is based on the keyboard clock and is synchronized with the board clock is generated. This clock is used in an **always** block to collect serial data bits and shift them into *ScanCode*. Another **always** block in this code, monitors completion of serial data collection and generates the state of the keys pressed and released. The details of these sections of the interface module are depicted in Figure 13.4 and Figure 13.5 respectively.

The first **always** statement of Figure 13.4 shows the generation of *Clock* that is equal to the keyboard clock and synchronized with the external system clock, *SYNclk*. In the **always** block that follows this block, after detection of the start-bit, on the rising edge of *Clock*, bit values are read from the keyboard data line (*KBdata*) and shifted into *ScanCode*.   The shifting continues for 8 bit counts. On the next clock after collecting 8 data bits is complete, *ScanRdy* is issued, and the collection process returns to its initial state of looking for the next start-bit.

```
...
reg Clock;

always @(posedge SYNclk) Clock = KBclk;

reg [3:0] BitCount;
reg StartBitDetected, ScanRdy;
reg [7:0] ScanCode;

always @(posedge Clock) begin
  if (ResetKB) begin
    BitCount = 0; StartBitDetected = 0;
  end else begin
    if (KBdata == 0 && StartBitDetected == 0) begin
      StartBitDetected = 1;
      ScanRdy = 0;
    end else if (StartBitDetected) begin
      if (BitCount < 8) begin
        BitCount = BitCount + 1;
        ScanCode = {KBdata, ScanCode[7:1]};
      end else begin
        StartBitDetected = 0;
        BitCount = 0;
        ScanRdy = 1;
      end
    end
  end
end
...
```

Figure 13.4 Serial Data Collection

The partial code of Figure 13.5 uses the two-bit *CompletionState* to keep track of the scan codes that have been generated. Starting in the initial state, when *ScanRdy* becomes **1** and F0 is on *ScanCode*, *CompletionState* becomes 1. This state is entered when a key is released and the F0 part of the Break code is transmitted. The next time *ScanRdy* is detected, the second part of the Break code (that is the same as Make) becomes available on *ScanCode*. In the following clock, the *KeyReleased* output becomes **1** and remains at this level for a complete clock period.

```
. . .
reg [1:0] CompletionState;
wire KeyReleased;

always @(posedge SYNclk) begin
  if (ResetKB) CompletionState = 0;

  else case (CompletionState)
    0: if (ScanCode == 8'hF0 && ScanRdy == 1) CompletionState = 1;
       else CompletionState = 0;
    1: if (ScanRdy == 1) CompletionState = 1;
       else CompletionState = 2;
    2: if (ScanRdy == 0) CompletionState = 2;
       else CompletionState = 3;
    3: CompletionState = 0;
  endcase

end

assign KeyReleased = CompletionState == 3 ? 1 : 0;

endmodule
```

Figure 13.5 Break State Recognition

13.3.2 ASCII Look-Up

The ASCII lookup part of our keyboard interface is a ROM of Quartus II megafunctions with 7 address lines and word length of 8 bits. Hexadecimal locations 0D through 66 of this ROM are defined to contain ASCII codes for scan codes that correspond to ROM addresses. This megafunction is defined to use the *KbASCII.mif* memory initialization file, a portion of which is shown in Figure 13.6. The *ScanCode* output of Figure 13.3 connects to the address input of this ROM, and ASCII codes corresponding to input addresses appear on its output.

```
DEPTH = 128;
WIDTH = 8;
ADDRESS_RADIX = HEX;
DATA_RADIX = DEC;
% Keyboard Scan Code to ASCII %
CONTENT
BEGIN
% Set 2:    ASCII    ;         Key  Char   %
%------+---------------+------------------%
0D    :    09    ; %   16   Tab   %
0E    :    96    ; %   1     `    %
11    :    0     ; %   60   Alt   %
12    :    0     ; %   44   Shift %
14    :    0     ; %   58   Ctrl  %
15    :    81    ; %   17   Q     %
16    :    49    ; %   2    1     %
1A    :    90    ; %   46   Z     %
1B    :    83    ; %   32   S     %
1C    :    65    ; %   31   A     %
. . .
66    :    08    ; %   15   BS    %
END;
```

Figure 13.6 ASCII Conversion Memory Initialization File

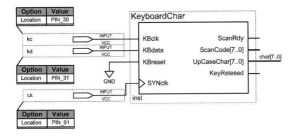

Figure 13.7 Prototyping Keyboard Character Generator

13.4 Keyboard Interface Prototyping

The Verilog module of Figure 13.3 and the ROM of Figure 13.6 are put together into the *KeyboardChar* schematic file. Testing this design is achieved by programming the FLEX 10K of UP2, assigning keyboard clock and data inputs to pins 30 and 31 (see Figure 6.37), and connecting the UP2 clock to its *SYNclk* input. A portion of the schematic of this prototype design is shown in Figure 13.7. With this settings, the ASCII code of the key pressed on the keyboard that is connected to the PS2 connecter of UP2 appears on the 8-bit *char* output of the diagram of Figure 13.7.

13.5 Summary

This chapter showed another interface design utilizing Verilog design entry as well as storage megafunctions of Quartus II. The design here illustrated how an interface could be designed to read keyboard characters. The knowledge of this basic peripheral is important for logic designers and students in this field.

14
Design of SAYEH Processor

This chapter shows design of a small computer in Verilog and implementation of it on UP2 using Quartus II. The CPU is SAYEH (Simple Architecture, Yet Enough Hardware) that has been designed for educational and benchmarking purposes. The design is simple, and follows the design strategy used for the multiplier of Chapter 1. We rely on the material of the chapter on computer architectures for providing the necessary background for understanding details of the hardware of SAYEH in this chapter.

14.1 CPU Description

The simple CPU example discussed here has a register file that is used for data processing instructions. The CPU has a 16-bit data bus and a 16-bit address bus. The processor has 8 and 16-bit instructions. Short instructions contain shadow instructions, which effectively pack two such instructions into a 16-bit word. Figure 14.1 shows SAYEH interface signals.

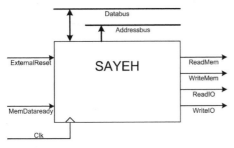

Figure 14.1 SAYEH Interface

14.1.1 CPU Components

SAYEH uses its register file for most of its data instructions. Addressing modes of this processor also take advantage of this structure. Because of this, the addressing hardware of SAYEH is a simple one and the register file output is used in address calculations.

SAYEH components that are used by its instructions include the standard registers such as the Program Counter, Instruction Register, the Arithmetic Logic Unit, and Status Register. In addition, this processor has a register file forming registers RO, R1, R2 and R3 as well as a Window Pointer that defines RO, R1, R2 and R3 within the register file. CPU components and a brief description of each are shown below.

- **PC:** Program Counter, 16 bits
- **R0, R1, R2, and R3:** General purpose registers part of the register file, 16 bits
- **Reg File:** The general purpose registers form a window of 4 in a register file of 8 registers
- **WP:** Window Pointer points to the register file to define RO, R1, R2 and R3, 3 bits
- **IR:** Instruction Register that is loaded with a 16-bit, an 8-bit, or two 8-bit instructions, 16 bits
- **ALU:** The ALU that can AND, OR, NOT, Shift, Compare, Add, Subtract and Multiply its inputs, 16 bit operands
- **Z flag:** Becomes **1** when the ALU output is **0**
- **C flag:** Becomes **1** when the ALU has a carry output

14.1.2 SAYEH Instructions

The general format of 8-bit and 16-bit SAYEH instructions is shown in Figure 14.2. The 16-bit instructions have the *Immediate* field and the 8-bit instructions do not. The *OPCODE* filed is a 4-bit code that specifies the type of instruction. The *Left* and *Right* fields are two bit codes selecting *R0* through *R3* for source and/or destination of an instruction. Usually, *Left* is used for destination and *Right* for source. The *Immediate* filed is used for immediate data, or if two 8-bit instructions are packed, it is used for the second instruction.

15 12	11 10	09 08	07 00
OPCODE	*Left*	*Right*	*Immediate*

Figure 14.2 SAYEH Instruction Format

Our processor has a total of 29 instructions as shown in Table 14.1. Instructions with *I* immediate field are 16-bit instructions and the rest are 8-bit instructions. Instructions that use the *Destination* and *Source* fields (designated by *D* and *S* in the table of instruction set) have an opcode that is limited to 4 bits. Instructions that do not require specification of source and destination registers use these fields as opcode extensions. Because of this, our processor has room for extending its instruction set beyond what is shown in Table 14.1. In addition to *nop*, hex code 0F is used as filler for the right most 8-bits of a 16-bit word that only contains an 8-bit instruction in its 8 left-most bits.

Table 14.1 Instruction Set of SAYEH

Instruction Mnemonic and Definition		Bits 15:0	RTL notation *:comments or condition*
nop	No operation	0000-00-00	No operation
hlt	Halt	0000-00-01	Halt, fetching stops
szf	Set zero flag	0000-00-10	Z <= '1'
czf	Clr zero flag	0000-00-11	Z <= '0'
scf	Set carry flag	0000-01-00	C <= '1'
ccf	Clr carry flag	0000-01-01	C <= '0'
cwp	Clr Window pointer	0000-01-10	WP <= "000"
mvr	Move Register	0001-D-S	$R_D <= R_S$
lda	Load Addressed	0010-D-S	$R_D <= (R_S)$
sta	Store Addressed	0011-D-S	$(R_D) <= R_S$
inp	Input from port	0100-D-S	In from port R_S and write to R_D
oup	Output to port	0101-D-S	Out to port R_D from R_S
and	AND Registers	0110-D-S	$R_D <= R_D \& R_S$
orr	OR Registers	0111-D-S	$R_D <= R_D \mid R_S$
not	NOT Register	1000-D-S	$R_D <= \sim R_S$
shl	Shift Left	1001-D-S	$R_D <= sla\ R_S$
shr	Shift Right	1010-D-S	$R_D <= sra\ R_S$
add	Add Registers	1011-D-S	$R_D <= R_D + R_S + C$
sub	Subtract Registers	1100-D-S	$R_D <= R_D - R_S - C$
mul	Multiply Registers	1101-D-S	$R_D <= R_D * R_S$ *:8-bit multiplication*
cmp	Compare	1110-D-S	R_D, R_S *(if equal:Z=1; if $R_D<R_S$: C=1)*
mil	Move Immediate Low	1111-D-00-I	$R_{DL} <= \{8'bZ, I\}$
mih	Move Immediate High	1111-D-01-I	$R_{DH} <= \{I, 8'bZ\}$
spc	Save PC	1111-D-10-I	$R_D <= PC + I$
jpa	Jump Addressed	1111-D-11-I	$PC <= R_D + I$
jpr	Jump Relative	0000-01-11-I	$PC <= PC + I$
brz	Branch if Zero	0000-10-00-I	$PC <= PC + I$ *:if Z is 1*
brc	Branch if Carry	0000-10-01-I	$PC <= PC + I$ *:if C is 1*
awp	Add window pointer	0000-10-10-I	WP <= WP + I

In the instruction set, addressed locations in the memory are indicated by enclosing the address in a set of parenthesis. When these instructions are executed, the processor issues *ReadMem* or *WriteMem* signals to the memory. When input and output instructions (*inp, oup*) are executed, SAYEH issues *ReadIO* or *WriteIO* signals to its IO devices.

14.1.3 SAYEH Datapath

The datapath of SAYEH is shown in Figure 14.3. Main components and their lower level structures are listed below.

1. Addressing Unit

       a. PC (Program Counter)

       b. Address Logic

2. IR (Instruction Register)
3. WP (Window Pointer)
4. Register File

       a. Decoder1(Left)

       b. Decoder2 (Right)

5. ALU (Arithmetic Unit)
6. Flags

As shown in Figure 14.3, components are either hardwired or connected through three-state busses. Component inputs with multiple sources, such as the right hand side input of ALU, use three-state buses. Three-state busses in this structure are *Databus* and *OpndBus*. Names shown on component interconnections are used in the Verilog description of the processor.

    In this figure, signals that are in italic are control signals issued by the controller. These signals control register clocking, logic unit operations and placement of data in busses.

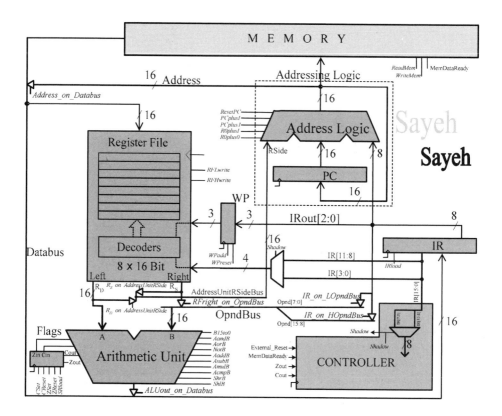

Figure 14.3 SAYEH Datapath

14.1.4 Datapath Components

Figure 14.4 shows the hierarchical structure of SAYEH components. The processor has a *datapath* and a *controller*. *Datapath* components are *Addressing Unit, Instruction Register, Window Pointer, Register File, Arithmetic Unit,* and the *Flags* register. The *Addressing Unit* is further partitioned into the *Program Counter* and *Address Logic*.

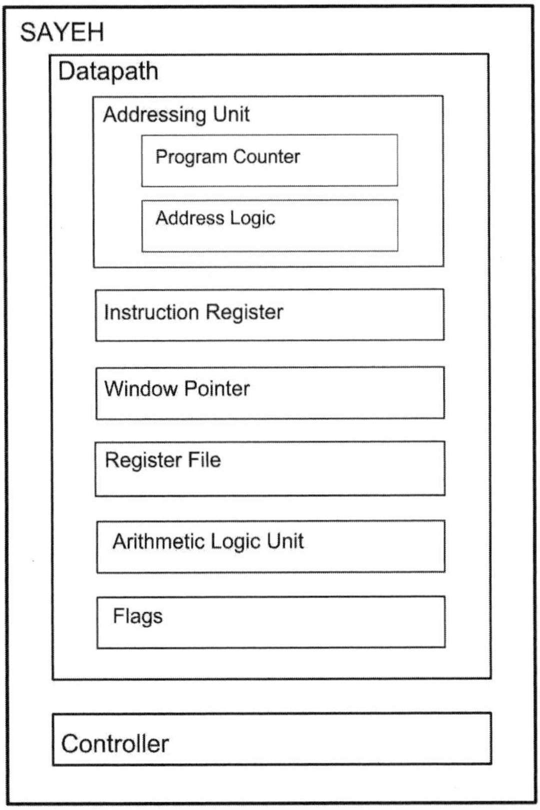

Figure 14.4 Sayeh Hierarchical Structure

The *Addressing Logic* is a combinational circuit that is capable of adding its inputs to generate a 16-bit output that forms the address for the processor memory. The *Program Counter* and *Instruction Register* are 16-bit registers. The *Register File* is a two-port memory and a file of 8 16-bit registers. The *Window Pointer* is a 3-bit register that is used as the base of the *Register File*. Specific registers for read and write (*R0, R1, R2* or *R3*) in the *Register File* are selected by its 4-bit input bus coming from the *Instruction Register*. Two bits are used to select a source register and other two bits select the destination register.

When the Window Pointer is enabled, it adds its 3-bit input to its current input. The *Flags* register is a 2-bit register that saves the flag outputs of the *Arithmetic Unit*. The *Arithmetic Unit* is a 16-bit arithmetic and logic unit that has the functions shown in Table 14.2. A 9-bit input selects the function of the ALU shown in this table. This code is provided by the processor controller.

Table 14.2 ALU Operations

Mnemonic	Description	Code
B15to0H	Place B on the output	1000000000
AandBH	Place A and B on the output	0100000000
AorBH	Place A or B on the output	0010000000
notBH	Place not B on the output	0001000000
shlBH	One bit shift to left	0000100000
shrBH	One bit shift to right	0000010000
AaddBH	Place A + B on the output	0000001000
AsubBH	Place A - B on the output	0000000100
AmulBH	Place A * B on the output	0000000010
AcmpBH	Z = 1 if A = B; C = 1 if A < B	0000000001

Controller of SAYEH has eleven states for reset, fetch, decode, execute, and halt operations. Signals generated by the controller control logic unit operations and register clocking in the datapath.

SAYEH sequential data components and its controller are triggered on the falling edge of the main system clock. Control signals remain active after one falling edge through the next. This duration allows for propagation of signals through the busses and logic units in the datapath.

14.2 SAYEH Verilog Description

SAYEH is described according to the hierarchical structure of Figure 14.4. Data components are described separately, and then wired to form the datapath. Controller is described in a single Verilog module. In the complete SAYEH description, the datapath and controller are wired together.

14.2.1 Data Components

Combinational and sequential SAYEH data components are described here. The combinational ones are like the ALU that perform arithmetic and logical operations. The function of such units is controlled by the controller. The sequential components are clocked with the negative edge of the main CPU clock. These components have functionalities like loading and resetting that are controlled by the controller.

```
module AddressingUnit (
    Rside, Iside, Address, clk, ResetPC, PCplusI, PCplus1, RplusI, Rplus0, PCenable);
input [15:0] Rside;
input [7:0] Iside;
input ResetPC, PCplusI, PCplus1, RplusI, Rplus0, PCenable;
input clk;
output [15:0] Address;
wire [15:0] PCout;
    ProgramCounter PC (Address, PCenable, clk, PCout);
    AddressLogic AL (PCout, Rside, Iside, Address, ResetPC, PCplusI, PCplus1, RplusI, Rplus0);
Endmodule
```

Figure 14.5 *AddressingUnit* **Verilog Code**

```
module ProgramCounter (in, enable, clk, out);
input [15:0] in;
input enable, clk;
output [15:0] out;
reg [15:0] out;

    always @ (negedge clk)  if (enable) out = in;

endmodule
```

Figure 14.6 *ProgramCounter* **Verilog Code**

```
module AddressLogic (
    PCside, Rside, Iside, ALout, ResetPC, PCplusI, PCplus1, RplusI, Rplus0);
input [15:0] PCside, Rside;
input [7:0] Iside;
input ResetPC, PCplusI, PCplus1, RplusI, Rplus0;
output [15:0] ALout;
reg [15:0] ALout;

    always @ (PCside or Rside or Iside or ResetPC or PCplusI or PCplus1 or RplusI or Rplus0)
        case ({ResetPC, PCplusI, PCplus1, RplusI, Rplus0})
            5'b10000: ALout = 0;
            5'b01000: ALout = PCside + Iside;
            5'b00100: ALout = PCside + 1;
            5'b00010: ALout = Rside + Iside;
            5'b00001: ALout = Rside;
            default: ALout = PCside;
        endcase

endmodule
```

Figure 14.7 *AddressLogic* **Verilog Code**

```
`define B15to0H 10'b1000000000
`define AandBH 10'b0100000000
`define AorBH  10'b0010000000
`define notBH  10'b0001000000
`define shlBH  10'b0000100000
`define shrBH  10'b0000010000
`define AaddBH 10'b0000001000
`define AsubBH 10'b0000000100
`define AmulBH 10'b0000000010
`define AcmpBH 10'b0000000001

module ArithmeticUnit ( A, B,
       B15to0, AandB, AorB, notB, shlB, shrB, AaddB, AsubB, AmulB, AcmpB, aluout,
       cin, zout, cout);
input [15:0] A, B;
input B15to0, AandB, AorB, notB, shlB, shrB, AaddB, AsubB, AmulB, AcmpB;
input cin;
output [15:0] aluout;
output zout, cout;
reg [15:0] aluout;
reg zout, cout;

   always @( A or B or B15to0 or AandB or AorB or notB or
             shlB or shrB or AaddB or AsubB  or AmulB  or AcmpB or cin)
   begin
     zout = 0; cout = 0; aluout = 0;

     case ({B15to0, AandB, AorB, notB, shlB, shrB, AaddB, AsubB, AmulB, AcmpB})
         `B15to0H:aluout = B;
         `AandBH: aluout = A & B;
         `AorBH:  aluout = A | B;
         `notBH:  aluout = ~B;
         `shlBH:  aluout = {B[15:0], B[0]};
         `shrBH:  aluout = {B[15], B[15:1]};
         `AaddBH: {cout, aluout} = A + B + cin;
         `AsubBH: {cout, aluout} = A - B - cin;
         `AmulBH: aluout = A[7:0] * B[7:0];
         `AcmpBH: begin
            aluout = A;
            if (A> B) cout = 1; else cout = 0;
            if (A==B) zout = 1; else zout = 0;
         end
         default: aluout = 0;
     endcase

     if (aluout == 0) zout = 1'b1;
   end

endmodule
```

Figure 14.8 *ArithmeticUnit* Verilog Code

Addressing Unit. The Addressing Unit, shown in Figure 14.5, consists of the Program Counter and Address Logic. The Program Counter is a simple register with enabling and resetting mechanisms, while the Address Logic is a small arithmetic unit that performs adding and incrementing for calculating PC or memory addresses.

This unit has a 16-bit input coming from the Register File, an 8-bit input from the Instruction Register, and a 16-bit address output. Control signals of the Addressing Unit are *ResetPC, PCplusI, PCplus1, RplusI, Rplus0,* and *PCenable.* These control signals select what goes on the output of this unit. Shown in Figure 14.6 is the Verilog code of the Program Counter. The Address Logic of Figure 14.7 uses control signal inputs of the Addressing Unit to generate input data to the Program Counter via the *PCout* of Figure 14.5.

Arithmetic Unit. The ALU of SAYEH is shown in Figure 14.8. For readability, control input codes of this unit are defined according to their function. For example, the select input that causes the ALU to perform the add operation is 0000001000, and it is defined as *AaddBH.* Control inputs of this unit are *B15to0, AandB, AorB, notB, shlB, shrB, AaddB, AsubB, AmulB* and *AcmpB* that select its various operations. In order to insure that no unwanted latches are made, all ALU outputs are set to their inactive values at the beginning of the **always** statement of its Verilog code. In a **case**-statement in this code, *aluout* and its flags outputs are set according to the selected control input of the ALU.

Instruction Register. SAYEH Instruction Register is shown in Figure 14.9. This unit is a 16-bit register with an active high load-enable input. As shown the only control input of the *InstructionRegister* module is *IRload.*

Register File. Figure 14.10 shows the Verilog code of SAYEH Register File. This is a two-port memory with a moving window pointer. For reading from the memory, the base of the window pointer (*Base*) is added to the left and right addresses (*Laddress* and *Raddress*) and memory words are read on appropriate left and right outputs (*Lout* and *Rout*). Writing into the memory is done in the location pointed by its left address (left is used for instruction destinations). The *RFLwrite* and *RFHwrite* control signals decide whether a write is done to the low order or the high order bits of the Register File. If both these signals are active, writing is done in a 16-bit word addressed by *Laddress* plus *Base.*

```
module InstrunctionRegister (in, IRload, clk, out);
input [15:0] in;
input IRload, clk;
output [15:0] out;
reg [15:0] out;

   always @(negedge clk)   if (IRload == 1) out <= in;
endmodule
```

Figure 14.9 *InstructionRegister* **Verilog Code**

```
module RegisterFile (in, clk, Laddr, Raddr, Base, RFLwrite, RFHwrite, Lout, Rout);

input [15:0] in;
input clk, RFLwrite, RFHwrite;
input [1:0] Laddr, Raddr;
input [2:0] Base;
output [15:0] Lout, Rout;

reg [15:0] MemoryFile [0:7];

wire [2:0] Laddress = Base + Laddr;
wire [2:0] Raddress = Base + Raddr;

assign Lout = MemoryFile [Laddress];
assign Rout = MemoryFile [Raddress];

reg [15:0] TempReg;

   always @(negedge clk) begin
     TempReg = MemoryFile [Laddress];
     if (RFLwrite) TempReg [7:0] = in [7:0];
     if (RFHwrite) TempReg [15:8] = in [15:8];
     MemoryFile [Laddress] = TempReg;
   end

endmodule
```

Figure 14.10 *RegisterFile* Verilog Code

14.2.2 SAYEH Datapath

Figure 14.11 shows the datapath of SAYEH module. This module specifies component instantiations and bussing structure of the CPU according to the diagram of Figure 14.3. Inputs of this module are the processor's data and address busses, as well as control signals that are provided by the controller of the CPU. Control signals shown in the Data Path module are routed to the data components that are instantiated here or the internal buses that are specified in this module.

Following the declarations, the Data Path module instantiates Addressing Unit, Arithmetic Unit, Register File, Instruction Register, Status Register, and the Window Pointer. Control signals that are inputs of *DataPath* are passed from this module to the data components via their port connections. For example, *ResetPC* that is an input of *DataPath* and a control signal of *AddressingUnit* appears on the port list of *AddressingUnit* in its instantiation statement.

The part that follows module instantiations makes bus assignments to the internal buses of this module. For example, assignment of the output of *ArithmeticUnit* to *Databus* that is controlled by *ALU_on_Databus* is done by an **assign** statement with a right hand side that is a conditional expression. Note the assignment of 16'bZZZZZZZZZZZZZZZZ to *Databus* when none of control signals of this bus are active.

```
module DataPath (
  clk, Databus, Addressbus,
  ResetPC, PCplusI, PCplus1, RplusI, Rplus0,
  Rs_on_AddressUnitRSide, Rd_on_AddressUnitRSide, EnablePC,
  B15to0, AandB, AorB, notB, shlB, shrB, AaddB, AsubB, AmulB, AcmpB, RFLwrite, RFHwrite,
  WPreset, WPadd, IRload, SRload, Address_on_Databus, ALU_on_Databus,
  IR_on_LOpndBus, IR_on_HOpndBus, RFright_on_OpndBus,
  Cset, Creset, Zset, Zreset, Shadow, Instruction, Cout, Zout );
input clk;
inout [15:0] Databus;
output [15:0] Addressbus, Instruction;
output Cout, Zout;
input
  ResetPC, PCplusI, PCplus1, RplusI, Rplus0,
  Rs_on_AddressUnitRSide, Rd_on_AddressUnitRSide, EnablePC,
  B15to0, AandB, AorB, notB, shlB, shrB, AaddB, AsubB, AmulB, AcmpB, RFLwrite, RFHwrite,
  WPreset, WPadd, IRload, SRload, Address_on_Databus, ALU_on_Databus,
  IR_on_LOpndBus, IR_on_HOpndBus, RFright_on_OpndBus,
  Cset, Creset, Zset, Zreset, Shadow;
wire [15:0] Right, Left, OpndBus, ALUout, IRout, Address, AddressUnitRSideBus;
wire SRCin, SRZin, SRZout, SRCout;
wire [2:0] WPout;
wire [1:0] Laddr, Raddr;
  AddressingUnit AU (AddressUnitRSideBus, IRout[7:0], Address, clk,
                     ResetPC, PCplusI, PCplus1, RplusI, Rplus0, EnablePC);
  ArithmeticUnit AL (Left, OpndBus, B15to0, AandB, AorB, notB, shlB, shrB,
                     AaddB, AsubB, AmulB, AcmpB, ALUout, SRCout, SRZin, SRCin);
  RegisterFile RF (Databus, clk, Laddr, Raddr, WPout, RFLwrite, RFHwrite, Left, Right);
  InstrunctionRegister IR (Databus, IRload, clk, IRout);
  StatusRegister SR (SRCin, SRZin, SRload, clk, Cset, Creset,
                     Zset, Zreset, SRCout, SRZout);
  WindowPointer WP (IRout[2:0], clk, WPreset, WPadd, WPout);

  assign AddressUnitRSideBus = (Rs_on_AddressUnitRSide) ? right :
                     (Rd_on_AddressUnitRSide) ? Left : 16'bZZZZZZZZZZZZZZZZ;
  assign Addressbus = Address;
  assign Databus = (Address_on_Databus) ? Address :
         (ALU_on_Databus) ? ALUout : 16'bZZZZZZZZZZZZZZZZ;
  assign OpndBus[07:0] = IR_on_LOpndBus == 1 ? IRout[7:0] : 8'bZZZZZZZZ;
  assign OpndBus[15:8] = IR_on_HOpndBus == 1 ? IRout[7:0] : 8'bZZZZZZZZ;
  assign OpndBus = RFright_on_OpndBus == 1 ? Right : 16'bZZZZZZZZZZZZZZZZ;

  assign Zout = SRZout;
  assign Cout = SRCout;
  assign Instruction = IRout[15:0];

  assign Laddr = (~Shadow) ? IRout[11:10] : IRout[3:2];
  assign Raddr = (~Shadow) ? IRout[09:08] : IRout[1:0];
endmodule
```

Figure 14.11 SAYEH *DataPath* Module

In the last part of the *DataPath* module, bits of *IR* that indicate source and destination registers to the Register File are placed on *Laddr* and *Raddr* inputs of this register. The *Shadow* signal that becomes **1** if a shadow instruction is being executed is used to select appropriate bits of the *IR* as source and destination addresses.

14.2.3 SAYEH Controller

The controller of SAYEH is a state machine with nine states that issues appropriate control signals to the Data Path. The controller uses the Huffman style of coding, in which the state machine has a large combinational part that is responsible for state transitions and issuing controller outputs. State transitions are done by setting next state values to the *Nstate*. Figure 14.12 shows a general outline of this controller. Various sections of this outline are discussed below.

Controller Ports. The instruction register output, ALU flags, and external control signals constitute the inputs of the controller. The outputs of the controller are 38 control signals going to the Data Path and a *Shadow* output that indicates that the controller is handling a shadow instruction. As shown in Figure 14.12, controller outputs are declared as **reg** and are assigned values in the combinational **always** block of the controller module.

Control States. A **parameter** declaration declares the eight states of the controller. States *reset* and *halt* are for the initial state of the machine and its halt state. In state *fetch* the machine begins fetching a 16-bit instruction that can include an 8-bit instruction and a shadow. State *memread* is entered while our controller is waiting for *memDataReady* signal from the memory indicating that its data is ready. Execution of instructions is performed in the *exec1* state. This state is entered from the *memread* state. The *lda* instruction that is not completed by the *exec1* state requires the additional state of *exec1lda* to complete its memory read. States *exec2* and *exec2lda* are like *exec1* and *exec1lda* except that they handle the shadow part of an instruction. The execute state of most instructions (*exec1* or *exec2*) increment the program counter while the instruction is being executed. However, certain instructions that use the address bus for their execution cannot increment *PC* while they are being executed. For these instructions, the *incpc* state increments the program counter.

Opcodes. Referring to Figure 14.12, instruction opcodes are declared as 4-bit parameters in the controller of SAYEH. These parameters are according to the processor's instruction set of Table 14.1.

State Declarations. As mentioned, the coding style the controller is according to the Huffman style of Figure 3.56 discussed in Section 3.3.4. The next state and present states, required by this style of coding, are declared in the controller of SAYEH as 4-bit registers, *Nstate* and *Pstate*.

```
module controller (
 ExternalReset, clk, ResetPC, PCplusl, PCplus1, Rplusl, Rplus0, . . . );
input
```

```
  ExternalReset, clk, . . .
output
  ResetPC, PCplusl, PCplus1, Rplusl, Rplus0, . . .
reg
  ResetPC, PCplusl, PCplus1, Rplusl, Rplus0, . . .
parameter [3:0]
  reset = 0,    halt = 1,       fetch = 2,      memread = 3,
  exec1 = 4,    exec2 = 5,      exec1lda = 6,  exec2lda = 7,      incpc   = 8;
parameter nop = 4'b0000;
parameter hlt = 4'b0001;
parameter szf = 4'b0010;
. . .
reg [3:0] Pstate, Nstate;

wire ShadowEn = ~(Instruction[7:0] == 8'b000011111)

  always @ (Instruction or Pstate or ExternalReset or Cflag or Zflag or memDataReady) begin
      ResetPC          = 1'b0;
      PCplusl          = 1'b0;
      PCplus1          = 1'b0;
      Rplusl           = 1'b0;
      Rplus0           = 1'b0;
      . . .
      case (Pstate)
        reset :
        . . .
        halt :
        . . .
        fetch :
        . . .
        memread :
        . . .
        exec1 :
        . . .
        exec1lda :
        . . .
      _ exec2 :
        . . .
        exec2lda :
        . . .
        incpc :
        . . .
        default:  Nstate = reset;
      endcase
  end

  always @ (negedge clk) Pstate = Nstate;

endmodule
```

Figure 14.12 SAYEH Controller General Outline

Shadow Instructions: The *ShadowEn* signal that is internal to the controller is set when the hex code 0F (this code indicates that the right-most bits are not used) is not found in the right-most eight bits of a 16-bit instruction. If this wire is **1** and execution of an 8-bit instruction is complete, the controller branches to *exec2* to execute the second half of the instruction before the next fetching begins.

Combinational Block. The combinational block of SAYEH controller has an **always** block that has a main **case** statement with case choices for every state of the machine. Transitions from one state to another and issuing control signals are performed in the case statement. At the beginning of the **always** statement, all control signals are set to their inactive values in order to avoid latches on these outputs.

```
always @ (Instruction, Pstate, ExternalReset, Cflag, Zflag) begin
  . . .
    case (Pstate)
    . . .
      exec1 :
        if (ExternalReset == 1'b1) Nstate = reset;
        else begin
          case (Instruction[15:12])
          . . .
            mvr : begin
              RFright_on_OpndBus = 1'b1;
              B15to0 = 1'b1;
              ALU_on_Databus = 1'b1;
              RFLwrite = 1'b1;
              RFHwrite = 1'b1;
              SRload = 1'b1;
              if (ShadowEn==1'b1)
                Nstate = exec2;
              else begin
                PCplus1 = 1'b1;
                EnablePC=1'b1;
                Nstate = fetch;
              end
            end

            lda : begin
              Rplus0 = 1'b1;
              Rs_on_AddressUnitRSide = 1'b1;
              ReadMem = 1'b1;
              Nstate = exec1lda;
            end
            . . .
          endcase
        end
    endcase
end
. . .
```

Figure 14.13 Instruction Execution

Sequential Block. The last part of the code outline of Figure 14.12 is the sequential **always** block f or clocking *Pstate* into *Nstate*. The control state register of SAYEH and all its data registers are falling edge trigger. Control signals issues by the controller remain active through the next falling edge of the system clock.

Instruction Execution. Figure 14.13 zooms on the combinational **always** statement of the *controller* **module** and shows the details of execution of *mvr* in the *exec1* state of the controller. Signals issued for the execution of this instruction are shown in this figure. This instruction reads a word from the right address of the Register File and writes it into its left address. The right and left (source and destination) addresses are provided in the Data Path by connections made from *IR* to the Register File.

```
always @ (Instruction, Pstate, ExternalReset, Cflag, Zflag) begin
  . . .
  case (Pstate)
    . . .
    exec1lda :
      if (ExternalReset == 1'b1)
        Nstate = reset;
      else begin
        if (memDataReady == 1'b0) begin
          Rplus0 = 1'b1;
          Rs_on_AddressUnitRSide = 1'b1;
          ReadMem = 1'b1;
          Nstate = exec1lda;
        end
        else begin
          RFLwrite = 1'b1;
          RFHwrite = 1'b1;
          if (ShadowEn==1'b1)
            Nstate = exec2;
          else begin
            PCplus1 = 1'b1;
            EnablePC=1'b1;
            Nstate = fetch;
          end
        end
      end
    . . .
  endcase
end
. . .
```

Figure 14.14 Memory Handshaking for *exec1lda*

The *RFright_on_OpndBus* control signal is issued to read the source register from *RegisterFile* onto *OpndBus*. Since this bus is the input of the ALU, the data on the ALU's right input (*B*) must pass through it to reach its output. For this purpose, the *B15to0* control input of ALU is issued. Once the data reaches the ALU output, it becomes available at the input of the Register File. Issuing *RFLwrite* and *RFHwrite* cause data to be written into the destination into *RegisterFile*.

The partial code of Figure 14.13 shows assignment of *exec2* to *Nstate* if the instruction we are executing has a shadow. Otherwise, signals for incrementing the Program Counter are issued and the next state is set to *fetch*.

The execution discussed here applies to most SAYEH instructions. However, instructions that require memory access, e.g., *lda*, require an extra clock for reading the memory. The first part of the execution of *lda* is shown in Figure 14.13. As shown, for the execution of this instruction, the address is read from Register File and put on the address bus. At the same time, *ReadMem* is issued to initiate the memory read process.

The next state for execution of *lda* after *exec1* is *exec1lda* shown in the partial code of Figure 14.14. In this state, *ReadMem* continues to be issued and state remains in *exec1lda* until *memDataReady* becomes **1**. In this case, memory data that is available on *Databus* will be clocked into *RegisterFile* by issuing *RFLwrite* and *RFHwrite*.

Executions of other SAYEH instructions are similar to the examples we discussed. The complete Verilog code of SAYEH controller is over 800 lines and is included on the CD that accompanies this book.

14.2.4 Complete SAYEH Processor

The top-level Verilog code of SAYEH that is shown in Figure 14.15 consists of instantiation of *DataPath* and *controller* modules. In *Sayeh* module, control signal outputs of *controller* are wired to the similarly named signals of *DataPath*. The ports of the processor are according to the block diagram of Figure 14.1.

```
module Sayeh ( clk, ReadMem, WriteMem, ReadIO, WriteIO,
    Databus, Addressbus, ExternalReset, MemDataready);
input clk;
output ReadMem, WriteMem, ReadIO, WriteIO;
inout [15: 0] Databus;
output [15: 0] Addressbus;
input ExternalReset, MemDataready;

wire [15:0] Instruction;
wire esetPC, PCplusl, PCplus1, Rplusl, Rplus0,
    . . .
DataPath dp ( clk, Databus, Addressbus,
    ResetPC, PCplusl, PCplus1, Rplusl, Rplus0,  . . . );

controller ctrl ( ExternalReset, clk,
    ResetPC, PCplusl, PCplus1, Rplusl, Rplus0, . . . );

endmodule
```

Figure 14.15 SAYEH Top Level Description

14.3 SAYEH Testing

Because of the complexity of this design, it is best to test it with an HDL simulator and a high level testbench. Tools for generation and application of test data and monitoring and generation of output data are provided in HDL simulators. These tools together with ability to describe high level testbenches provide an efficient test and debugging environment for HDL based designs.

The testbench for SAYEH is shown in Figure 14.16. The use of external files for reading and writing test data are demonstrated by this example. As shown in this figure, *SayehRAM* that is a memory of 1024 16-bits words is declared in this testbench. The testbench reads test data that is the memory image of our processor in this file and when the test is completed contents of this memory are written into another external file. The input file is *SayehRAM.hex* and the output file is *OutputRAM.hex*. Contents of both files are in hexadecimal. 16-bit hexadecimal codes in these files represent memory data starting from location 0.

The first **initial** block is labeled *IOfiles*. This block opens the *OutputRAM.hex* output file for later writing and reads the contents of *SayehRAM.hex* into the declared *SayehRAM* memory. Reading the input file (memory image) is done by the **$readmemh** system task. This task expects data in the file to match the word length of the memory it is writing into.

An **always** block shown in *SayehTest* testbench generates a periodic signal on the circuit clock input.

The next procedural block shown in this testbench is an **initial** block that is labeled *RunCPU*. This block applies the resetting signal, runs the CPU for 370,000 nanoseconds, and when this time expires, it writes all 1024 words of *SayehRAM* into *OutputRAM.hex* external file. Note here that the **$fopen** statement in the *IOfiles* block made *memout* a file handler for the output file. The **$stop** statement in *RunCPU* block stops the simulation after the memory image has been written.

The **always** procedural block that is labeled *MemoryRead* handles reading data from *SayehRAM* when requested by the CPU. When *ReadMem* is issued by the CPU, the testbench issues *MemDataready* and places data from *SayehRAM* at the *Addressbus* location on *MemoryData*. At all other times, *MemoryData* bus is at the high-impedance state. This is done because *MemoryData* connects to *Databus* hat is a bi-directional bus.

The **always** block that appears next in Figure 14.16 handles writing data that appears on *Databus* into *SayehRAM*. This block has delays to allow signals from the CPU to stabilize.

This testbench allows for any SAYEH program to be loaded into the CPU memory and executed. Out testing of this processor consisted of an instruction based testing as well as several programs. For the instruction testing we applied independent instructions and monitored internal registers of SAYEH. For example, F205, that is the hex code for "*mil r2 05*", loads 05 into R2 of the Register File. Similarly, 0204 is the packing of two 8-bit instructions that set the zero and carry flags. An initial testing of a CPU requires verification of individual CPU instructions.

A more elaborate test program is discussed in the next section.

```
`timescale 1 ns /1 ns

module SayehTest ();

reg clk, ExternalReset, MemDataready;
reg [15:0] MemoryData;
wire [15:0] Databus, Addressbus;
wire ReadMem, WriteMem, ReadIO, WriteIO;

reg [15:0] SayehRAM [0:1023];

integer memout;
initial begin : IOfiles
   memout = $fopen ("OutputRAM.hex");
   $readmemh ("SayehRAM.hex", SayehRAM);
   clk = 0; ExternalReset = 0; MemDataready = 0;
   MemoryData = 16'bZ;
end

always #20 clk = ~clk;

integer i;
initial begin : RunCPU
   #05 ExternalReset = 1; #81 ExternalReset = 0;
   #370000;
   for (i=0; i<= 1023; i=i+1)
      $fdisplay (memout, "%h: %h", i, SayehRAM[i]);
   $stop;
end

always @(negedge clk) begin : MemoryRead
   if (ReadMem) begin
      #1 MemDataready = 1;
      MemoryData = SayehRAM [Addressbus];
   end else begin
      #1 MemDataready = 0;
      MemoryData = 16'hZZZZ;
   end
end

always @(negedge clk) begin : MemoryWrite
 #1 if (WriteMem) #1 SayehRAM [Addressbus] = Databus;
end

assign Databus = MemoryData;

Sayeh U1 (clk, ReadMem, WriteMem, ReadIO, WriteIO,
            Databus, Addressbus, ExternalReset, MemDataready);

endmodule
```

Figure 14.16 A Testbench for SAYEH

14.4 Sorting Test Program

Figure 14.17 shows a sorting program for SAYEH. This program reads data starting from the CPU memory and sorts them in descending order. The number of data item to sort is in location 768 and data begins in the next memory location. This sorting program uses two loops for the sorting to be done. When completed, the CPU is put into the halt state.

```
0000 mil r0 00     r0= 768       starting address in memory
0001 mih r0 03
0002 lda r1 r0     r1=           total number of elements
0003 awp 5
0004 mil r0 01     r5=1          for adding with index each time
0005 mih r0 00
0006 cwp
0006 add r1 r0     r1=           limit for final r4
.0007 mvr r2 r1
0008 awp 2
0009 sub r0 r3     r2=           limit for index r3
0009 cwp
000A mvr r3 r0     r3=           outer index
000A nop
000B cwp
000B cmp r3 r2                   check if outer index is equal to its limit
000C brz 19                      branch to 0025 if zero
000D awp 3
000E add r0 r2     r3=r3+1       increment outer index
000E mvr r1 r0     r4=r3         set inner index to outer index as initial
000F cwp
0010 awp 1
0011 cmp r3 r0                   check if inner index is equal to its limit
0012 brz 10                      branch to 0022 if zero
0013 awp 2
0014 lda r3 r0     r6=(r3)
0015 awp 1
0016 add r0 r1     r4=r4+r5      increment inner index
0016 lda r3 r0     r7=(r4)
0017 cmp r2 r3                   check if r6 is greater than r7
0018 brc 07                      branch to 001F if carry
0019 lda r1 r0     r5=(r4)       r5 as an temperary register
0019 sta r0 r2     (r4)=r6
001A cwp
001B awp 3
001C sta r0 r2     (r3)=r5
001D mil r2 01
001E mih r2 00     r5=1          for adding with index each time
001F cwp
0020 awp 5
0021 jpa r0 0E                   jump to 000F
0022 cwp
0023 awp 5
0024 jpa r0 0A                   jump to 000B
0025 hlt
```

Figure 14.17 Sorting Program for SAYEH

The program shown in Figure 14.17 is translated into its hexadecimal equivalent and is put in *SayehRAM.hex* file. As discussed in the previous section, SAYEH testbench reads instructions from this file and applies to the CPU.

14.5 FPGA Programming

The CPU described in this chapter has been programmed into the FLEX device of an Altera UP2. We used a RAM from Altera's megafunctions and configured it as a memory of 1024 16-bit words. The number of logic elements used by this CPU was 1,125, which is 30% of the available LEs. Memory bits used was 16,384, which is 44% of the available memory bits. This usage indicates that we can form a complete system with a keyboard and VGA output on a FLEX 10K of UP2.

14.6 Summary

This chapter showed design, testing and implementation of a complete CPU. This design put all that we have covered in this book into one package. The design is complete and typical of any large system with a complex controller and data path. Use of the synthesizable subset of Verilog for development of a design for FPGA programming was shown. On the other hand, utilization of behavioral constructs of Verilog was demonstrated in developing a testbench for our processor.